S0-BRT-591

Sincerely,
Virginia Law Burns

TALL
ANNIE

Ana K. Clemenc
1888–1956
The Copper Miners' Heroine

To Tabby, my best friend.

TALL
ANNIE

by
Virginia Law Burns

Enterprise Press
Bath, Michigan

Enterprise Press
P.O. Box 108
Bath, MI 48808

Printed in the United States of America

Printing: 1 2 3 4 5 6 7 8 9 10
Year: 7 8 9 0 1 2 3 4 5 6

Library of Congress Cataloging in Publication Data

Burns, Virginia Law
Tall Annie.
Includes index.
1987   LC 87-80726
ISBN 0-9604726-3-0

Contents

Forward

Of the four sons and daughters of Michigan I've biographed so far, "Annie's" life has been the most difficult. Her school and church records were destroyed by fire; the first prime source I found was her marriage certificate.

I owe more than a few thanks to her youngest brother, Frank Klobuchar, the only close family survivor. With great patience and good nature, Frank answered my hundreds of questions by mail, telephone and personal interview. I could not have written this book without him.

I based Annie's childhood years, then, on Frank's recollections as well as the lives of other miner's children during the early 1890's. Mostly, the dramatizations are taken from facts, and where I have used exact quotations, the print is in italics.

Read in Good Health!

Virginia Law Burns
Haslett, Michigan
April, 1987

Preface

In the late 1970's on a trip to the Upper Michigan's Copper Harbor, my husband, Gerald, and I stopped for a quick coffee break in Calumet.

This delightful northern Michigan city was in the midst of a celebration of local history. Intrigued by a display of historic photos in a local photography shop, we decided to look more closely. A feature of the display was the Copper Strike of 1913. One picture stood out: that of a beautiful young woman carrying a flag and marching at the head of a column of strikers. That woman was identified only as "Tall Annie," and to find out who 'Tall Annie' was became the object of a mystery we were determined to solve.

Our research led us first to the article, "Big Annie and the 1913 Copper Strike" by Clarence Andrews, a former professor of Humanities at Michigan Technological University and an authority on Upper Peninsula history. We were later fortunate to make the acquaintance of Frank Klobuchar, Annie's younger brother who shared with us memories of childhood and his sister.

We learned that "Tall Annie" was Annie Clemenc, a woman who had played a key role in the copper strike.

When the Michigan Women's Studies Association began the development of the Michigan Women's Hall of Fame and Historical Center in 1980, the association adopted Annie Clemenc as one of our symbolic figures. Her likeness is one of three which appear on the Life Achievement Award presented to Michigan Women's Hall of Fame honorees. An early activity of MWSA was to organize a "Day of Tribute" to Annie and to commemorate a portrait of her by Andy Willis of the Miner's Art Group.

That portrait will hang in the Michigan Women's Hall of Fame.

We chose "Tall Annie" because she embodied those qualities of courage and commitment in which we believe. Annie, like many other women, had been lost to history until rediscovered.

This book will do much to let people know about a true Michigan heroine.

Gladys Beckwith
East Lansing, Michigan
March 1987

Acknowledgement

The author gratefully acknowledges the Beckwith's permission to use Andy Willis' painting as a significant and beautiful part of this biography. Dr. and Mrs. Beckwith have been most helpful and supportive the two years "Annie" was in the typewriter.

Thanks also, to Prof. Arthur Thurner, whose impeccably researched history of Michigan's Upper Peninsula saved me weeks of leg work. Dr. David Halkola gave encouragement and Mrs. Mary Kaefisch patiently dug up bits of history at the Calumet Library. Peggy Germain, Clarence Monette and Judith Dow are new friends, whose enthusiasm and assistance kept me inspired.

Lastly, my thanks to Monica Berger of Dover Publications, who graciously gave permission to use the quotation from James Oppenheimer's poem, "Bread and Roses."

CHAPTER 1

GEORGE AND MARY'S FIRST CHILD

Their eyes shining with love, George and Mary marveled at their first-born, a daughter, just minutes old.

"She's tall already," George teased, using his tender voice. He rubbed, feather-light, a large, work-roughened finger across the baby's forehead.

"How could she not be tall, with us for parents?" Mary murmured. She looked up at her husband, towering beside the painted metal bed where she and the child lay. Mary herself had often laughed about her own height when George was courting her in Slovenia. "I wouldn't have given you a second glance if you were short," she had said. Many South Slavs physically resembled their close neighbors, the Austrians—they were small statured with fair hair and light-colored eyes.

Now Mary's long, brown hair spread loosely over her shoulders. She would have liked to comb, pull and fasten it back into a bun, but she was exhausted. A neighborhood midwife attended the birth, and after admiring and swaddling the well-formed baby, had bustled off to the kitchen to make tea.

"Do you still want to name her 'Ana'?" George asked, pulling a small wooden chair to the bedside. The question could have gone unsaid, for George knew that once his young wife had made a decision, she rarely vacillated.

Mary's eyes fluttered closed and the corner of her full lips turned up.

George leaned over and whispered loudly, "Hey—don't drift off! Tea's ready and honey cakes, too."

But Mary Klobuchar, her task finished, had slipped into a deep sleep. George scraped back the chair and stood a long moment, thinking how suddenly a man becomes the head of a family.

Now his meager laborer's wages from the Michigan copper mines would have to support three.

Other miners manage; so can I, George thought. He turned to the kitchen and caught a glimpse of the calendar on which Mary had been crossing off the days. This day's date, March 2, 1888, was not marked.

*　　*　　*

Slightly more than a week after Ana's birth, and six days after Benjamin Harrison took the oath of office for presidency of the United States, the country was hurled into a wave of concern and excitement. The Samoan Islands were the topic of a dispatch reporting that a German warship had torpedoed an American war vessel. American newspapers were outspoken in their hostile opinions, and Draft-age men, most of them having no idea where Samoa was, became uneasy. But truth finally overtook rumor. A monsoon and tidal wave had sunk warships of Britain and Germany, as well as those of our nation. By the time Ana was a year old, the three countries had signed an agreement to share equally in the prosperous trade of the tropical islands.

An episode which would eventually influence Ana Klobuchar was unfolding in London in 1888. Jane Addams, an outstanding advocate for social reform (and who later became a Nobel Prize winner) visited England. There, she carefully studied operations of a large home where workers helped immigrants "settle in" to the ways of a new culture.

Jane brought the ideas home to Chicago and implemented them in a more democratic form. She called the homes, "Neighborhood Centers" and established programs ranging from day nurseries to college courses; they were

2

open to all. Years later, from the elegant "Opera House" stage in Calumet, Michigan, Jane Addams spoke for labor reform.

*   *   *

The cold, early spring day that brought Ana Klobuchar into the small world of Redjacket was to bring a year of change into the life of her parents and hundreds of other Slovenes. All had traveled half-way around the globe to find work and freedom in the new copper mines of Michigan's Keweenaw Peninsula. The Slovenes, as many of the other Balkan immigrants, were predominatly Catholic by religion and the church was an integral, undergirding support to them. Their priest headed a great family of often-times homesick, troubled, impoverished or grieving parishioners. The English-Latin-speaking Sacred Heart Catholic Church grew so crowded at every Mass that latecomers found themselves standing in the rear for the entire service.

Ana was probably baptized at Sacred Heart as an infant (all records were lost when this church burned.) That very year, 1888, the Slovene Delegation petitioned the old country for their own permanent priest, and the home church obliged by sending the good father, Joseph Zolokar.

Taking heart, the Slovene/Americans began a monthly money-raising project to build their own church. (Eventually there were seven ethnic Catholic churches in the Calumet area). The Calumet and Hecla Mine Company donated two thousand dollars and two city lots. Members of St. Joseph's Slovene Lodge reacted with the speed of lightning, and under the skilled hands of the immigrant carpenters and builders, work progressed with fervor.

"Isn't it grand we will have our own priest and new church, too!" Mary said. She tied a napkin around baby Ana's neck and slipped her into a sturdy, homemade wood highchair.

"I promised to help carry hod this week-end," George said, "Or do whatever the volunteer crew foreman needs."

Mary laughed. "Good. We'll say 'Hello' to you on our way to pick berries."

3

"Fine help she'll be," George said, tousling Ana's black hair, then helping himself to the stewed chicken and dumplings. "She hasn't started to crawl yet."

"Never you mind," Mary said. "She can get the thimbleberries closest to the ground and watch out for bears."

Ana, delighted to be part of the happiness, crowed and banged her spoon on the tin dish.

Playing or watching baseball games at Agassiz Park was another free pastime. The game had been around since the 1870's, but in early days, played only by professionals, whose managers, naturally, charged admission. Soon, it became widely popular on vacant lots, and thousands of youngsters learned to play during the town's beginning years. Ana, as a teenager, spent happy hours with younger sister Mary and brother George, practicing catching and batting.

"Don't worry. You can't hit home runs every time," George would tease. "Just whack the ball and run. With your long legs, you can stretch a single into a triple."

1888 also saw roller skating become a fad and a frustration, for there were few level, smooth surfaces to skate on. Then, a roller rink was built. When Mary and George took Ana for air in a hand-me-down wicker buggy, they would stroll past the rink smiling at the organ music, loud and cheerful, which accompanied the rumble of small wooden wheels—wheels which often went in directions not anticipated by the skater.

Billiards were a male-only pursuit. Weekdays, when Mary wheeled little Ana on downtown streets, she kept her pace brisk as she passed by the billiard halls, where beer smells, smoke and raucous conversation drifted out to the sidewalks. Women considered the game a low form of entertainment.

The tourist business, surprisingly, brought in seasonal but tidy profits to restauranteers and innkeepers. Large numbers of visitors came by boat and train to be taken on conducted tours of the copper mines. Tourists hired buggies and horses and trotted to the countryside for picnics, walked Lake Superior's shores and brought back pockets bulging with water-polished agates and "pretty stones."

Mary's outings with Ana frequently took them to the church building site. The cornerstone had been laid, with proper ceremonies, thanking all the donors. The Klobuchars were proud of the effort and gave as much as George could squeeze from his small wages. Ana's first winter came with the usual snow, bitter winds—and misfortune! Lake Superior gales demolished the skeleton structure of the new church.

Sad, but undaunted, the parishioners began another financial drive and by the time Ana had learned to walk the building was habitable.

CHAPTER 2

GROWING AND DREAMING
IN CALUMET

On a brilliant autumn day in 1890, Mary bent over her kitchen table, punching and turning bread dough. Flour motes drifted through a shaft of sunshine which slanted from a small window. She looked up as a neighbor rushed in the back door.

"Have you heard? The train was robbed! They got the whole C. & H. payroll!" The woman panted a little from her run across lots.

Mary stared, her hands deep in the spongy mass. No payroll! How would they buy food and pay the rent? George had worked a month for this pay and now it was gone to thieves!

"Do you know anything else?" Mary said in a pinched voice.

"No," the woman was already retreating. "I'm going to dress the babies and walk down to meet my Angelo when the shift ends. Maybe I can pick up some news on the way."

Ana watched, puzzled, as the long skirted figure swept away and the door banged. Clutching her rag doll, she ran to her mother and leaned against her leg.

Mary smiled down at her. "I put this bread in pans to rise, Ana, and we go for a walk, too."

By the time elevators had born the day shift miners to the surface, Calumet was buzzing. Passengers on the train who had witnessed the holdup had disembarked and were recounting their stories with gusto.

When George met his wife and daughter on the street, tin lunch pail under his arm, they became one family of many waiting for news.

At home, her brow creased with anxiety, Mary pulled Ana into her lap, settled them both comfortably in the chair and began to rock.

"Do you think the workers will get their pay soon, George?"

"Like I said, rumor has it the payroll was only seventy thousand dollars—not the usual million." George headed for the kitchen.

He poured hot water from a teakettle into a china bowl, and began to wash his face. "If that's so, the company will just wire Boston for more money—they sure got plenty there." He towelled his craggy features. "I heard there's a posse organizing and a couple of Sheriff Bowden's cronies have been at the scene all afternoon."

"I saw a few National Guard men at the station," Mary said. "What good are they?"

"Supposed to check all trains leaving here so the robbers can't escape by rail, I guess."

"Humph," Mary slid Ana to the floor. "A fast horse could take them anywhere in the woods around the county."

George emerged from the kitchen, scrubbed and grinning. "That's what posses are good for, Sugarplum," he said, giving his wife a hug. Swinging Ana up and cradling her in one long, strong arm, he picked a dill pickle from the dinner table, bit off a piece and offered the remainder to Ana. She wrinkled her nose and George popped the pickle into his mouth.

"Sheriff Bowden has already deputized a lot of fellows with good horses and they're out beating the bushes."

"But what if they don't catch the bandits?" Mary worried. "What about the workers' wages?" She brought a large tureen of stew from the stove.

"Like I said, C. and H. is rich enough to lose more than that. We'll be paid. Now let's eat!"

As George predicted, the mining company received a quick replacement payroll, and the sheriff's department continued its investigation. In a short time, thanks to a distinctive horseshoe print in the soft earth where the train had been stopped, the robbers were traced to an express messenger. He had been in charge of the payroll and had provided information to his rascally brother and friends, who actually did the dirty work. Ironically, before they could ship the loot out of the upper peninsula, the station master in Marquette and a cohort stole the trunk of money from the baggage room and buried it in an old kerosene can near a stream. They never revealed how they had discovered the stolen cash.

Within a few weeks, the conniving men, none of whom had been in trouble before, were tried and sentenced to five years in Marquette Prison. The five thousand dollars reward was split up between delighted investigating officers.

* * *

Just as the roller skating fad had engulfed Calumet, so did bicycling. Merchants couldn't keep stocked with the vehicles.

Hobbyists formed clubs, planned races and surprised themselves by riding the hundred miles to Marquette on the newly-invented macadam road.

By 1896, the bicycle craze had forced the Calumet-Red Jacket village officials to write new ordinances. The "safety bicycle" had replaced the awkward, bone-breaking high wheeler. This new bike sported wheels of the same size; thereafter and yearly, improvements like rubber tires, coaster brakes, cushion saddles and adjustable handle bars made it hard for some enthusiasts to contain their speeds, even though pedalers going faster than seven miles an hour were fined, if caught. None, however, in Ana's family owned a bicycle. George's wages could not be stretched for such a frippery and the children knew better than to ask their mother for money she brought in by cooking and washing.

But there were spectator sports that were free. The summer that Ana was eight, a daredevil balloonist came to town. His name was Hunter, and there must have been a lot of children, like Ana and her 4-year old sister, Mary, whose breath caught in their throats as Hunter poised a thousand feet above Calumet in the reed basket and JUMPED! It was their first sight of a parachute opening and landing.

By this time, 1896-97, Ana was attending one of the several good schools operated and paid for, of course, by the school district from local taxes. The mining company (C. & H.) had built schools on company land, hired the teachers, and leased back the property to the school board. Ana received an excellent education. Her classmates were largely immigrant children who had to learn English along with their writing and arithmetic. In this regard, the teachers did superb work with these children, handcrafting visual aids and using pupils whose first language was English, as tutors.

In Michigan's Upper Peninsula, winter says "Hello" about two weeks earlier than it does in the lower peninsula. By February, the Keeweenaw area usually has twelve and often sixteen feet of snow on the level. "Snowjumping" became "the game". Klobuchar girls, as other northern children, took turns leaping from the second story bedroom window into the snowbanks. Little Mary squirmed valiantly through the drifts, far deeper than she was tall, toward the cleared sidewalk. She never made it by herself. Ana always had to help; they would run to the back door, quickly brush the snow from their woolens, and dash up the stairs for another fall.

Some years, their father cleared rocks and raked a level spot in the back yard early in the winter. He would pump water by hand and flood a small ice skating area for his girls. Downtown, there was a square which had been watered with a fire hose, and ice skating there was graced with the bright strains of the Fifth Regiment Band.

Snow removal in Calumet begins in November and goes on and on.

In 1897, the year baby George was born, crude wooden boxes, perhaps fifteen feet long by six feet wide, were attached to long, homemade skiis, and drawn by a team of sturdy horses. City workers then shoveled snow into the "snowbox". It was beastly slow but it was all Calumet had. During the January and February blizzards, business, schools, society doings, railroads, trolleys . . . everything came to a silent halt and waited. Waited for the winds and snows to abate. Townspeople then dug themselves out of their homes and businesses and hoped that the snow crews would soon reach their street. In March, people spoke longingly of the "breakup" or thaws of spring. "Bye, bye, cracked bones", they said. "So long, slippery sidewalks."

Ana, now nine, and Mary five, could hardly wait to take little George out for air each day. This was a real help to their mother. "I don't mind hard work", she would say, "but the Lord needs to make a longer day if I'm ever to get it all done!"

A miner's pay would not cover the expenses of a large family, and wives supplemented the income the only ways that were open to them . . . cooking, washing, and cleaning. Mary Klobuchar had become known as an excellent cook; she catered weddings, funerals, parties and various festive celebrations. This, in addition to abundantly caring for her children, house and husband and two or three bachelor boarders.

Women of these turn-of-the-century-years also spent much time on church activities. In a mining town, charity was a necessity. Immigrants arrived daily and there were no professional greeting societies—only the ethnic clubs, which were largely religious. Mary Klobuchar, a mother three times by the year 1897, found she had a gift for midwifery. Giving birth in hospitals was not popular with women, perhaps with good reason. Institution postpartum infections were common and sometimes fatal for both mother and child, and home births seemed to be safer. At any rate, Mary earned a little extra this way. She was learned, too, in medicinal herbs and home remedies. The Klobuchar children often woke at night to the sound of worried adult voices—perhaps a crying infant. They would hear cabinet doors opening and closing, smell the pungent herbal odors, sometimes spicy, sometimes acrid. When they got up in the morning, there might be a note on the kitchen table from their mother saying where she was and who was sick; that Ana should make breakfast and get the younger children dressed. This kind of neighborly help and concern not only was charitable, but imperative. Many families could not afford the monthly dollar fee for company hospital care. A dollar could be a third to a half of a miner's daily pay.

Fire was an ever-present dread in Ana's early days. All the rental and company houses were of frame construction; there were even log cabins available. These were more primitive than the two-story wooden houses, but cheaper. Stores, churches—all structures were of wood. It was cheap, plentiful and took relatively unskilled labor to erect a passable dwelling. The Klobuchars were a lucky family. Their various rented homes were never damaged by fire.

When Michigan began to quarry sandstone from the nearby Portage area, C. & H. officials took to using this stone for whatever new building they needed. Mine administrators built palatial homes from this handsome, fireproof local stone.

Ever the paternal employer, C. & H. Mines completed a fantastic sandstone library in 1898. Its design was beautiful. Even those who had a hard time saying anything nice about the "Mine Bigwigs" conceded that the structure was a magnificent addition to a north woods community. From the day Ana learned to read she enjoyed the privileges of this free, public library. She would browse through the foreign language books on the first floor, challenging herself to find a printed word in a Slovene book which she had heard her parents use when they were with other Slovenes. Her first friendship with fiction bloomed here.

In later times, Ana would climb the stairs to the reading room, sliding her hand over the darkly shining handcrafted woodwork. There, as reference books looked down from the shelves, Ana lost herself in the newspaper reports and biographies of famous labor reform leaders. The handsome grandfather clock chimed away segments of time, and Ana dreamed of meeting these leaders—persons like Eugene

11

Debs, Mother Jones, Jane Addams and John L. Lewis.

Probably the most surprising feature of the library was the bathhouse in the basement. It too was free, had separate facilities for male and female, and even furnished towels at no charge. The Klobuchars, surviving on the edge of poverty, used the baths as often as Mary could gather up her brood and walk them over there. What joy in using all the hot water and soap she needed without having to heat it on a wood stove in a copper boiler. And wonders! No towels to wash afterwards!

Ana especially appreciated the baths. As eldest daughter, she inherited the job of Mother's Helper. They stirred the largest items on wash-day, with a stout stick. They rinsed and wrung the dead-heavy sheets by positioning themselves at opposite ends of the bedding, and twisting it. "At least we won't have so many towels to do now," Ana said to her mother. "That scrub board takes the skin off my knuckles and just when it grows back, it's time to wash clothes again!"

Mary smiled and glanced at her own chapped hands. "Maybe when you grow up, you marry a rich man. No more scrubbing and going to the free baths."

"I'm going to be a labor leader, like Mother Jones. I won't stay home long enough for a wash day! I'll just travel and give speeches."

"You young people!" Mary said, clucking. "Going to change the world in a day!" Then, seeing that Ana was very serious about labor injustices, softened.

"So have your dreams, little one. Someone has to speak up. My mother taught me not to settle for other people's ideas if I think they're wrong. But you finish grammar school—promise me that, Ana."

"Sure, Mama. My teacher says education is the way to a better life. But I want to make Papa's job better for him, too."

The year that Ana turned ten years old, the usual January Thaw became the "February Hot Spell". The temperatures for several days rose to sixty five degrees! Mary Klobuchar sent Ana outside with little Mary and George.

"It's so nice, you all go for a sled ride while I finish dinner," she said.

Ana was pleased. It would be fun to play outdoors with Mary and George instead of helping her mother in the kitchen. Ana lined a low, wooden box with a small blanket, set it atop their homemade sled, and loaded baby George into the box. The girls took turns pulling the sled; George clung with a death grip to the sides. Ana would have preferred to do all the pulling. She enjoyed vigorous exercise and she liked being in control. Slender and tall, she dominated her sister by several inches.

"See!" Mary scolded. "You pulled too hard when you started across the curb." The hearty tug had slipped out the sled from under George's box and he lay sprawled on his back, whimpering as the wet snow seeped under his collar.

"All right," Ana said, settling George again and tucking the blanket around his legs, "So you pull awhile." She handed Mary the rope. "I'll walk behind." They joined other children who lived on the block and played until the sun, quick and pale, sank behind the gigantic mine buildings.

That night, snugged down in her feather bed, Ana could hear one set of sleigh bells after another tinkle past her window. Townspeople had rented sleighs from the numerous liveries and formed parties to take advantage of the bizarre shift in temperature. Normally it would have been seventy-five degrees colder.

*　　*　　*

Because of the incredible amount of wealth copper was generating, there were means to make Calumet (which included Redjacket) a modern, almost cosmopolitan city. Considering the great number of foreign languages being spoken on every street corner, it was already, in a sense, cosmopolitan. Saturday night in Redjacket was a milling, crowded, noisy but generally well-mannered affair.

When the township built a theater, Calumet residents were thrilled and proud. They called it "The Opera House" and it opened the year the century turned. The structure was

13

of stone and every bit as grand as the library. First-rate talent actively competed for bookings there. John Philip Sousa and his band, the lovely singer Lillian Russell, and Jane Addams, spokeswoman for the downtrodden emigrants, were among the notables who performed at the beautifully appointed theater, along with various Shakespearean Companies. It was many years before Ana or any of her family could afford the price of admission (twenty-five cents to a dollar).

Ana was there, however, as an adolescent, when Eugene V. Debs addressed the audience from the sixty-feet-high stage. Debs was a prominent Socialist, whose ideas were particularly attractive to all types of miners and manual laborers. By the time Ana would become the symbol of striking copper miners in 1913, Debs would have run for three presidential elections on the Socialist ticket. The last campaign he directed from a prison cell, because he had given a speech protesting the ongoing World War. He received nearly a million votes and lost to Democrat Woodrow Wilson. Debs used the three years he was jailed to write the book, *Walls and Bars*, in which he described prison problems and conditions.

CHAPTER 3

FALLING IN LOVE

Father Pakiz had said Mass—The Feast of the Immaculate Conception—and the whole Slovene religious community had buttoned coats, buckled galoshes and headed out into a miserably cold December day. They climbed into their black, open carriages, or walked toward home. Ana, now a leggy fourteen years old, heard someone shout. Turning toward the sound, she saw people staring and gesturing at the church belfry.

Faint trails of smoke floated away from the steeple. Surely that could not be fire, she thought. We were just inside and there was no indication. Yet the unthinkable had happened! Ana and her family huddled together with other church members and watched, grieving, from across the street, as five fire departments fought the blaze and lost. St. Joseph's Church—destroyed again! Poor Father Pakiz escaped with nothing but his life, and barely that. Numbed by the loss of his church and home, the Father was hardly aware of the kindly hands leading him to a warm house, to fresh clothing and a meal hot from the oven. A parishioner sent word that the cross had fallen, unharmed, on a lot nearby. They would use it, of course, Father Pakiz promised himself, in the construction of another church. He knew his parish would not sit around wondering what to do.

"From prosperity to ashes," the congregation mourned. Of the three hundred and fifty families and four hundred single men, none felt the grief more than the Klobuchars. In

January, all kinds of committees met and out of them came plans for a new structure . . . this one to be of sandstone. The Bishop and Treasurer of the Diocese protested the great outlay of money.

"So it costs a lot," George Klobuchar said the spring of 1903, "But we've lost two churches from fire . . . all our work and money and records gone up in smoke."

"It's crazy to make the same mistake again. We need a more fireproof material," Mary agreed. "We can sacrifice some, and I know the societies will earn enough for the furnishings."

"I hope the building committee can get going on it soon," George complained. "The Italians have been good to us, letting us use their church, too, but we're used to having our own place to worship."

So the Klobuchars and many other determined Slovenians went prepared to the April meeting. Some came into Redjacket's five short streets on electric streetcars which connected all the neighboring settlements. Trams carrying others also clattered on schedule, the fifteen miles from the southern cities of Houghton and Hancock. Many hitched up horses and buggies; locals carefully made their way over the slippery wooden sidewalks. (There were a few prosperous lot owners who had cement sidewalks.)

If parishioners came from the direction of the railroad station, they could feel as well as hear the eerie underground noises they were all accustomed to . . . the rumbling and clanking of the iron ropes and rollers moving down and up the sloping mine shafts. The worshipers passed few Slovene businesses. It would be two decades before Balkan butcher shops, mercantiles and saloons appeared. Taverns, no matter what the ethnic ownership, always prospered. In addition to having a mind-boggling array of spirits, these establishments always sported nickel slot machines, live music and paid dancing girls.

By the time the church members had filed out the doors of their borrowed church this damp, chill April evening, they had pledged the cost of a new building . . . forty-six thousand dollars! It was an astonishing amount. Families

16

had promised to give eighty-five dollars and single men, fifty. For George Klobuchar, that was thirty-four days pay.

In June, the Slovenes watched with rightful satisfaction as their leaders, smiling broadly, broke ground with gleaming shovels. In August the cornerstone was laid for the new St. Joseph's Catholic Church. It said, "Let this church be a monument to our Spiritual Endeavor; and let our posterity view it as an emblem, symbolizing the 'Freedom of Worship.'"

Ana and George spent many pleasant hours picking Thimbleberries

Ana was one of the scores of young women of the parish to work at all manner of schemes to raise money to pay for an altar. Young men also put on various kinds of entertainment to bring in funds for an organ. A helpful and much-needed spin-off of this men's group was that they became a benevolent society dedicated to the welfare of other laboring Slovenes. Different ethnic groups had little jokes about them. "When a dozen or more Slovenes meet, they either discuss building a church or organizing a lodge." In 1905, when sixteen-year-old Ana went with her parents to the first services of St. Joseph's, they worshipped in the basement.

"It's good to have our own place," Ana whispered to her mother as they settled themselves on borrowed chairs. Mary smiled up at Ana, now taller than she by several inches. She switched baby Joseph, born the past year, to the other knee, and glanced down the aisle to see how her husband was managing little George. Mary was twelve years old, craning her neck and scanning the room for her girlfriends. They waved or made faces at her in recognition. Ana was more interested in the young men, and one youth in particular. She noticed that Joseph Clemenc, rangy and handsome, turned his dark-eyed gaze toward her more and more often.

He was quiet, amiable and tall—so tall that Ana had to tilt her head to look into his face. Ana found that her mind often was filled with thoughts of Joe—would he like her new hairdo, or the calico dress she made for herself? Did he find her intelligent as well as attractive?

*   *   *

It was to be another four years until the big beautiful church sat completed and furnished. Then how the bands marched through Redjacket!

Just as smartly, paraded all the Catholic Lodges in Calumet and the Slovene Societies in the area. When the organ's magnificent voice soared in its salute to God, members' hearts swelled. They had a right to be proud.

The edifice had cost more than a hundred thousand dollars and was paid for! At the same time as the great Slovene church effort was reaching its peak, a feisty, well-known old lady called "Mother Jones" was marching with other reformers to President Teddy Roosevelt's home on Long Island, New York. They were protesting child labor. Ordinarily, Mother Jones confined her considerable energies and talent to problems of iron workers. But the plight of youngsters caught in the web of inhumane textile factory owners touched her deeply. She strode briskly, a plump, dark little bird, holding herself very erect as she gripped a protest sign and chanted slogans with the protesters. Ana would soon read about Mother Jones' activities and a decade later, march with the famous woman in the streets of Redjacket.

Another outspoken critic of labor management was Jane Addams. Miss Addams' work in an east coast settlement house gave her first-hand knowledge of the terrible realities in being an immigrant worker in the New Industrial World. Her book, *Democracy and Social Ethics*, published in 1902, pointed out that "Progress must always come through the individual who varies from the type, and has sufficient energy to express this variation. He first holds a higher conception than held by the mass of his fellows of what is righteous under given conditions, and expresses this conviction in conduct." Jane Addams had described Mother Jones perfectly, and to a lesser extent, the Ana Clemenc of 1913. Personal problems, however, were to keep Ana from making a lifetime crusade of labor reform.

* * *

The next year, 1905, Frank, the last Klobuchar child, was born. Ana had always been valuable help in the household and now was needed more than ever.

"I couldn't manage without you, dear Ana," her mother said one late afternoon. Mary was cooking for a funeral dinner and the house smelled of smoked sausage and sweet cabbage rolls.

Ana grinned. "I guess now is as good a time as any," she said, putting her arm around her mother, who had just taken a pie from the oven.

"Time for what?" Mary said, setting the pie on the kitchen window sill to cool.

"To tell you that Joe and I have decided to get married," Ana said.

"I knew it! I knew it! Just when I get you trained good, you leave!" She threw her arms around her tall daughter and the two of them danced and whirled around the overheated kitchen.

*   *   *

Ana was eighteen years old, Joseph a few years older, also of Slovene descent. The wedding, of course, was held at the lovely, new St. Joseph's church. Joe's parents, George and Aloysiae, stood beside them, along with Mary Klobuchar. All signed the certificate as witnesses. It was May, 1906, and the newly-wedded couple settled in Red-jacket, for Joe Clemenc was a miner. Ana delighted in displaying an immaculate house.

"I'm a real workhorse," she often said. Her proud husband agreed. "And she cooks real fine, too," he'd smile, patting his lean stomach. Joseph liked a hearty laugh; it helped a miner through the dark places in his work life, he said. Joe had a beautiful, rich, baritone voice and a natural bent for singing harmony. The tavern where Joe hoisted a few drinks, (or more), usually rang with song, when Joe obliged a request for a solo or joined a quartet. At church, less talented choir members depended on his strong, clear voice to lead the way on new songs.

CHAPTER 4

PRELUDE TO STRIFE

A day and a half's ride south by train from Redjacket, was Chicago. There, the novelist, Upton Sinclair, had swung headlong into the ideals of Socialism. This year, 1906, his book, *The Jungle,* burst into the labor reform camp. Sinclair got himself hired in a Chicago meat packing company. His subsequent exposé of the gruesome practices and unsanitary conditions, along with the awful poverty of the workers, was a big influence in the quick flurry of Congressional investigations. Out of them came the National Pure Food and Drug Act and the Meat Inspection Act. Socialism was beginning to have a real impact upon the lives of United States citizens.

*　　*　　*

Ana's first years of marriage brought no children. She had time left over from her duties as a wife, landlady with boarders, and church worker. She had the opportunity to read intensively and note the helplessness of the working men in her mining town. It was true that the C. & H. Mine did much for the laborers: a hospital, a library, bathhouse, and donations of land and cash were paternalistic gestures that did make life more bearable. But the company chose for the workers what it saw to be good for them. The company kept wages so low that employees seldom got out of debt. There were few disability or death provisions. The many widows and women with crippled men worked themselves to an early grave trying to keep their children fed and housed. Ana cursed the company inwardly, even as she prayed for families that had met tragedy in the mines.

Slovenes as a nationality have traditionally been seen by other cultures as having a fierce national pride and a strong dislike of outside interference in their lives by government or employers. Ana was as true to the Slovene temperament as she was different in appearance to the average Slovene. She had grown to be six feet tall . . . some said two inches more than that. For all her height, she was slender, and if not as graceful as the more petite, it went unnoticed as she sped good naturedly through the dailiness of long winters and brief summers.

History has few facts of Ana's early life . . . her church records burned in the two St. Joseph's Church fires, and the Redjacket Grammar School she attended also went up in smoke, before her marriage to Joe. Ana probably dropped out of school to work after she graduated from the Eighth Grade, as was the custom with low income families. Yet Ana, and those like her who could not pursue their education further, were literate. In Houghton County, where Ana grew up, the literacy rate, in 1910, for all persons over ten years of age, was 94.3% For foreign born, it was 89.6%! The library was heavily used, especially the reading room where the area's seven foreign language newspapers were available.

People scrimped to save money to attend the "Opera House." Class- conscious Bostonians, many of whom were becoming very rich from Michigan copper, would have rolled their eyes if they discovered that the dirty miners, who made their wealth possible, helped time pass faster in a twelve-hour workday, by discussing the Shakespearean play they had seen the previous evening.

*   *   *

Ana and Joe's boarders lounged at the dining room table, eating dessert and having a last cup of tea. Snow fell fast and thick in Redjacket and the town lay quiet.

Snow muffled the raucous protests, even, of a nervous dog somewhere down the street. Ana cleared the table and poured the last drops of tea into Joe's cup.

"One of my friends from Hancock tells me Big Louis Moilanen is going to leave farming and buy a bar in town," a young miner said to no one in particular.

"Huh, Fella," said another miner, "You're behind times. Louis's already got a pub in Hancock."

"I bet he had to remodel the place," laughed another. "How tall is he, anyway?"

"Eight feet, two or three," the first boarder said. "Must weigh about four hundred pounds."

"The poor man," Ana said, settling herself at the table with the others. "Imagine how hard it must be for him to live, bumping his head in doorways and not being able to sit in regular chairs, and everyone always staring."

Joe spoke. "Yeah . . . I heard he has a sign in his tavern that says he doesn't cater to drunks or women," he said. "A brawl started one night and Louis put an end to it by grabbing the collar of each of the fellas and hoisting them straight to the ceiling!"

Sad, freakish Louis Moilanen also made a living for a short time with the Ringling Brothers Circus as a giant. He tried mining, but his inordinate size made his movements awkward and subsequently, dangerous to himself. Before he died at age twenty eight, he had gone back to farming.

*   *   *

George and Mary and the three younger children— George Jr., Joseph and Frank, were the remaining Klobuchar family in 1910. Mary wed very young and left Michigan. Ana settled within walking distance, and often dropped in to maintain the traditional Slovene family closeness. Her parents had never rented a company house. "C. & H. has too much control over my life, as it is," George would say when friends asked why he didn't take the easier and cheaper housing offered by the company. "I'll be damned if I'll do any more business with them than I have to."

The frame house on 6th Street, behind the Redjacket Fire Station, was their home for some years. Young Frank was five that year, and though there was no money for entertainment, they made their own fun. They ice skated; whittled their own skiis. In summer, they camped on Lake Superior beaches and picked berries. And they picnicked. How they picnicked! Slovene Lodges and religious groups had earned the reputation for organizing the biggest and most delectable picnics in the whole north country. A local businessman would loan his wagon and best team of horses for the occasion, (perhaps for an invitation?) and the fun began right after church on Sunday. Laughing, joshing menfolk handed up large food baskets to the wagon loaders, who sniffed and threatened to raid the provisions en route. The women, brandishing long wooden serving spoons, would warn, "If you want broke knuckles, you just try."

Picnics were the Klobuchar boys' favorite pastime, too, but since they were all school age, there weren't as many free hours. George, being the oldest, went his own way. Frank and Joe were constantly together. Their mother, starting that year, found part-time jobs for the three of them as soon as school let out in May. "It's as much to keep them out of mischief as it is for money," she laughed, telling Ana. Yet that summer, Joe was caught stealing apples from a neighbor's yard, arrested and fined twenty dollars, which his mother paid. Justice seems a little strained in this case. Twenty dollars fed a large family for more than two weeks.

After that, Mary saw that the boys were far too busy to get into trouble. They screened coal ashes for the president of the local bank, saving out chunks that could be re-burned. On Saturday, at Turk's Coffee House, where the brew bubbled in little, long-handled brass pots, they cleaned and tidied up for fifty cents a month.

They picked rags, bottles, copper and lead wire to sell to the town junk man. Mary let them keep these pennies, which they spent on candy or silent movies.

Berry picking in late summer was a day-long endeavor. With only lunch to break the monotony, the boys could fill several five-pound lard pails. They peddled raspberries and thimbleberries for thirty-five cents a pail, and what they couldn't sell, Mary made into juice or pies.

Frank and Joe sorted lettuce from the trash bins behind the neighborhood grocery store. It if were winter, Mary saved the best for her family and the lesser leaves went to the rabbits. The boys cared for these animals, which were used for meat . . . a fact that Frank was never able to stomach. He became a vegetarian on days the rabbits turned up in stew. "I could never get their little faces out of my mind," he said, recalling those days from an 80-year old memory.

If the boys had any free time, and they thought their mother wouldn't find out, they mooched around Gekus's Fruit and Candy Store. The two Gekus boys were their pals, and on lucky days, Mr. Gekus would show them what sweeping, mopping or carrying he wanted done, and their reward would be some overripe fruit or a little bag of candy.

Once, at school, a student broke a rule. The teacher had not seen the transgression, but sensing that Joe Klobuchar knew who the culprit was, she descended upon him.

"All right, Joe," she demanded, "You always know where trouble starts. Who did it?"

Joe looked up at the formidable figure, now an enemy. "I don't know, Ma'am."

"Yes, you do!" she said, grabbing Joe's new Mackinaw jacket. He had not yet had time to take it off since recess.

Joe refused to talk. Infuriated, the teacher shook Joe until his jacket tore. Her anger spent, she returned to her desk. Joe remained silent.

That afternoon, the teacher was surprised to see an agitated Mary Klobuchar stride through the classroom door.

"My Joe came home from school saying you tore his jacket trying to make him tell on another kid." Mary's eyes flashed and she drew nearer to the teacher's desk. "Here's his Mackinaw. You sew it up now, and we'll wait."

She tossed the heavy, woolen garment at the stunned woman, stalked to a front-row desk and settled herself. The teacher sewed.

Little Frank was no angel, either. He once got his pants set afire because his brother-in-law, Joe Clemenc, came upon him stealing matches from Ana's cupboard. Joe snatched a broom and whacked poor Frankie, who had, unfortunately, already stowed the big wooden matches in his back pocket. The blow ignited his pants, and Frank ended up smoldering in the kitchen sink.

When he was about nine years old, Frank appeared in a concert with the school choir at Electric Park. It was spring, and he remembered all his life, the pride he felt, as he stood tall, wearing his best shirt and wool pants, and sang "The Bluebird."

The Klobuchars waved and applauded wildly at its conclusion.

"See, Joe," Ana said teasing, "You're not the only singer in the family."

Her husband smiled and clapped louder. He wished music could play a larger part in his own life.

The Klobuchar boys were not worked any harder than the average miner's sons of the day. Mary was anything but cold-hearted. Yet she was a responsible mother, and with her husband having to work 12 hour shifts, much of the child-raising fell to her.

Mary, herself, worked continually. She was in particular demand at weddings and other church feasts as a caterer. Frank and Joe got to tag along and stuff themselves with pop and cake, which they never had at home.

Also harboring a sweet tooth was auburn-haired George, Jr. He was alone in his bright coloring; Sister Mary and brothers, Joe and Frankie, had light brown hair. All, however, looked at the world through eyes of a hazel hue.

Young George's disposition was the gentlest of the Klobuchars. He was quiet-spoken; sometimes his quick wit was surprising. He avoided the scrapes that seemed to seek out his younger brothers. This elder son loved to accompany his father on long walks to the outskirts of town. Quick-stepping to keep pace with the tall, big-boned figure beside him, George cheerfully endured as his father whistled the same tune, endlessly. The melodic line never varied, through the years. Folks along the route would smile and say, "Here comes George. He don't talk much but you always know when he's passing by."

But the winter of 1910 brought tragedy. All three boys contracted diphtheria. Mary and George took turns holding their sons' heads over a smoky oak fire in the hope of reducing painful swollen throats. Whether the procedure had any medicinal properties is doubtful. Joe and Frank slowly lost their fevers, their throats opened and strength returned. George died, in spite of attention from the company hospital doctor, a man named Penny.

George and Mary were devastated; Ana was inconsolable. "George should not have died!" she cried. "Dr. Penny is an incompetent. He should be barred from practicing!"

Those around her knew that it was little use to try to dissuade Ana. She did not rest until she had presented enough evidence to the Medical Society to prove that the doctor was, indeed, responsible for her brother's death. Penny left town, never to practice in the Keweenaw area again. But it didn't bring back her darling George. Ana sat in vigil, with her family, their hearts aching. They wept as friends opened their arms and murmured not only in Slovenian, but Finnish, Italian, Croation, German, Polish, Lithuanian, Hungarian and the rich dialects of Wales and Ireland. Ana wept bitterly. Frank, only five years old, raised himself on tiptoe for a last look into the casket. He had never really known his big brother, but Ana carefully preserved her memories of George. She had an enlargement made of George's photograph, and kept it on view for as long as she lived.

*   *   *

Perhaps this medical skirmish was the beginning of Ana's brief entrance into the battles of social justice. Discontent among the miners was growing, nearly all of whom were immigrants. They had been encouraged by Copper Management, to come to Michigan because of the need for a large labor supply. Keweenaw labor had gotten themselves organized with help from the Western Federation of Miners and it was true that Michigan miners' wages compared favorably with some other American mine workers. But it was often said that Keweenaw mine operators preferred foreign workers because the language barrier gave the company an upper hand. Ana, miner's daughter, miner's wife, saw much in a miner's day that was unjust, unsafe and demeaning. With her lusty spirit of independence and action, she was prepared for the anger that came to a boil and spilled over the sides of the miner's pot of discontent.

July 23 of 1913 had come.

CHAPTER 5

STRIKE!

Ana would have liked to let herself go and be part of the celebration that was readying in the streets and vacant lots of Redjacket. The band music, carried on the warm summer winds, brought a smile to her lips as she hurried through the ever- present housework that Wednesday. A Firemen's Convention had started off the day with a parade and music; now a carnival company, with practiced precision, fastened together sideshow booths and rides.

Ana's mind was not on the festivities as she prepared the day's meals for her husband and boarders. She was thinking about the union position on working conditions and pay. It was two days beyond the deadline the union had given management to respond to their request for a conference. Something had to happen soon.

She kindled a fire in the black iron stove; the night's dampness left as the room warmed. Today, the miners, all fifteen thousand of them, were not going to report for work at the mine gates. They would strike!

Her throat tightened with excitement at the bold plan. They would have to pray that the Western Federation of Mine Workers would support them in spite of W.F.M.'s caution that the copper miners should not rush into a confrontation.

Joseph shuffled into the kitchen, yawning and tucking a loose cotton shirt inside his dark trousers.

"Smells good," he said. He pulled out a wooden, spindle-backed chair and sat to lace his boots.

Ana finished browning bacon slices, then sliced potatoes into a frying pan. Her thoughts slipped from the union to company management.

"Joe, I wonder how MacNaughton and his men will take this strike. Do you think the company will treat you like they handled the miners back in the '72 strike?"

"Who's to tell, Ana? That was a long time ago. You have as good an idea about the way the company operates and the miners feel as any of us."

Ana's dark eyes took on a pensive look.

"That I do. It's going to be a fight all the way." She took a deep breath; picked a handful of eggs from the wooden ice box. "Will you tell the men breakfast will be ready in five minutes?"

*   *   *

George Klobuchar always worked the day shift at the Redjacket mine for C. & H. This mine and two others were unionized. The two mines abstaining were small and on the fringes of the area. Oddly, a majority of workers at the C. & H. Mines had voted against a strike, feeling that the Western Federation of Mines and the American Federation of Labor might not keep their promises of financial and moral backing. All the other unionized mines, however, voted to walk out if the mine managers did not reply.

The deadline came and went. The companies did not respond.

*   *   *

As Ana strode toward the center of Redjacket, this July 23, she wondered what was happening at the Palestra Auditorium in the adjoining village of Laurium. She knew that all the active union miners would be there receiving last-minutes instructions from union executives. What had started out to be a proper, lawful procession from the Laurium meeting place became a roaring deviation, a running mass. By the time it reached the north side of the Calumet Street Mine, the mob was ready to use weapons—weapons that lay discarded on the ground—iron bolts, gas pipes, rocks and sticks.

The mines' security chief, anticipating disorder, had asked for and gotten volunteers from within the working force to act as deputies. These unfortunates hardly had a chance in the wave of strikers, who rapidly and violently ejected them from the mine buildings. Underground workers found themselves stepping out of elevators into a barrage of rocks, clubs and fists in the face. Blood flowed; some of the injuries were so grievous that victims were bedridden for a month or more. It was a rampage stemming from the men's long-felt, stifled anger. Anger at the dismal poverty their families endured and their own powerlessness in their jobs. Yet, the rage was not fed by union leaders (who had continually spoken for peaceful bargaining), but by simple mob psychology.

*   *   *

While Ana had been hurrying through her chores that morning, James MacNaughton, General Manager of the C. & H. Mines, was not having a second cup of coffee at his spacious Calumet Avenue home. He had answered the telephone that morning at seven. The voice on the other end was that of a mine manager saying that the walkout had started. Annoyed and a little surprised, Mac-Naughton was in his office within the hour. Not long after, the Houghton County Sheriff deputized two-hundred-fifty volunteers. They sat tight, then, knowing that there was a decisional meeting going on in the Palestra. Still, they were not prepared for the rioting that followed. Helpless and angry, the mine security chief followed the incensed miners from shaft to engine house to boiler. Redjacket had no police department as such, only the sherriff's greatly outnumbered deputized helpers. In two hours, the miners had attacked workers in fourteen C. & H. Mines and had worked their way to the mine next door to MacNaughton's home. Hands in his pockets, the chief watched from his office windows and seethed. Strikers were ousting engine house workers. Unused to his new role of observer, Mac-Naughton nervously jotted down times and events as he viewed them. He was in constant telephone contact with his managers and knew every happening of consequence. Witnessing the riots was an onion of a different skin. It stung. He hated not being in control. Another manager phoned, saying that strikers had shouted to night workers that "they would never come up alive" if they went underground.

"The men at Shaft #5 have been threatened with death, too," the manager went on. "The strikers are ugly, Mr. Mac-Naughton, and we have families to think of, as well as our own skins."

"Go home, then," MacNaughton said. He turned to an aide. "Get me fifty or sixty deputies. Arm them with axe handles. I'm going to Number 5."

Within minutes, a half-hundred men, clutching new axe handles, assembled outside the general manager's office. MacNaughton walked into their midst and the group moved off to the Number 5 shaft. There they found the pump-men, fearful, and calling out to the deputies for protection. MacNaughton dismissed the workers, assigned them body-guards, and the remainder of the group headed for a rescue station the company had speedily put together.

"Not much doing here," a nurse said. "I heard rioting is worse over at the Redjacket Complex."

MacNaughton pulled off his small, rimless spectacles and polished them with a linen handkerchief.

"Well, they must be getting hungry." He gazed about him in the twilight, hearing the distant shouts and catcalls. "Maybe they'll go home now. That's where I'm headed." There was more matter-of-factness in his manner than he felt.

*　　*　　*

At Redjacket, Ana saw with incredulous eyes the quick retreat of visiting firemen and carnival paraphernalia when word came of the outbreak. As it swept toward town, streets cleared; frightened citizens pulled paper window shades and locked doors. No one knew what to expect. Ana heard the roar of violence before she could see the swarm of dark forms moving from the Tamarack Mine to the engine house in Redjacket. Staying on the fringe of the crowd, shouting as loudly and crudely as any man, Ana became part of the greatest labor conflict the Michigan Copper Range had ever seen.

Activity slowed and stopped by nightfall, but before dawn the next morning, Thursday, there were an estimated 250 strikers assembled at Number 2 Mine. In the darkness, strikers patrolled the entrances and exits of the enormous cavern, intimidating anyone who even looked like a working man.

The company's security man swore in more volunteers. Fighting went on the entire day with surprisingly little damage to property, except for broken windows.

But any man who tried to work was attacked, and some who were not even mine employees fell wounded in the meleé.

Ana and Joseph left their bed early that next morning and hurried through a breakfast of oatmeal and tea. Ana had drawn a shawl around her shoulders in the chill of the peninsula's dawn; Joe wore a light woolen shirt, patched many times but still bright and clean. Downtown, in contrast to the previous morning, miners were everywhere. Disappointment swept over Ana, as she saw more and more men headed for the mines.

Leaving Joseph's side, she ran toward a fellow Slovene. Anger replaced chagrin as she struggled to keep her voice steady.

"Peter, don't go back. We need you. We must hang together."

The miner halted, looking up at Ana. Feeling her intensity, he stepped back, glancing uneasily about him.

"I know, Ana, I know. But there's another baby on the way. I just can't afford to lose my wages."

Suddenly, near the Redjacket mine entrance, a scuffle

broke out. Turning, Ana saw a line of pickets, those miners more ready to pursue the cause, trading blows with those less dedicated. Joseph dashed toward the action, Ana close behind.

As the day wore on, Ana's long, cotton dress wrinkled and drooped, and her heavy dark hair fell from its pinned nest. Finally, exhausted from the stressful and often violent encounters, Ana trudged home to prepare dinner for her boarders.

*　　*　　*

That same night, strikers gathered at the union hall and learned that the rumors were true that Sheriff Jim Cruse had telegraphed Michigan's Governor Ferris asking for the state's entire peace force to be sent to Calumet.

"And the governor," continued the union speaker, "thinks we're more dangerous than the Mexicans because he's sending the whole blamed militia up here, instead of south to the border, and that includes three brass bands!"

Amid the laughter that followed, someone cried, "Maybe he thinks that if the National Guard can't break us with guns and bayonets, they can music us to death."

Before the meeting ended that Thursday night, active unionists completed plans to eliminate the picket line, which they rightly reasoned led to confrontations and fighting. Instead, they would have orderly parades through town every day.

"We must impress the non-striking workers that we mean to use peaceful methods to encourage them not to work until the companies negotiate," the speaker concluded.

Ana, her cheeks bright with excitement, jumped to her feet. Joe Clemenc and her father glanced at each other, not entirely startled by her action.

"Let me carry our country's flag," she cried, forgetting her fatigue from being on the streets all day. "I'll wave it every step of the way, no matter how far we march!" Her brown eyes glowed. " We'll show the companies we mean business. This is not a feudal system and we're not serfs!"

*   *   *

On Friday, July 25, the third day of the strike, Ana and Joseph, again up before dawn, dressed in fresh clothes. Ana hooked the last button on her high, black shoes and within moments, the couple was striding the few blocks to the Italian Hall where strikers and their families were gathering. The paraders lined up, three and four abreast, in the street. At the edge of the column, wives carried babies, older children held hands with the younger ones; all stood ready to accompany the miners.

In front of the line, waited the union officials. They and everyone else, wore their "Sunday Best" clothes, and at the very head of march was Ana Klobuchar Clemenc. This event was to earn for her the affectionate nickname, 'Tall Annie', for anyone seeing the statuesque figure of the young woman was impressed with her bearing and the cheerful determination showing in her face. The flag she carried displayed 48 stars sewn on silk, and was so large it required a two-inch thick staff.

Annie's heart beat doubletime. Her Balkan heritage provided a heightened intensity of the moment; she felt like a racehorse at the starting gate.

From the corner of her eye, she saw the lead union official give a signal, and with one last look behind, she flashed a proud smile and took the first long strides into history. Through the streets of Redjacket, past Blue Jacket and Yellow Jacket mines, whose buildings resembled gigantic boxes of squares and rectangles balanced atop one another—their roofs sharp triangles to better shed winter's heavy snow . . . past all the mine tailing piles formed from C. & H. operations . . . the paraders marched in the July heat. They had no band—this line was quiet, orderly. "See," their faces and posture announced, "We have dignity. We have rights as new Americans. Our grievances are real and deserve to be heard."

*   *   *

In bed that night, Annie eased her aching shoulders and arms away from a lump in the old mattress. Joe was already asleep. Annie lay very still and thought about the day's events. On Sunday, she mused, I shall wear my white dress. Maybe tie some ribbons to the flagstaff.

This young, second-generation immigrant woman could not have dreamed what she was marching into.

CHAPTER 6

LET THE PARADE BEGIN

Sunday morning saw the strikers and their families streaming toward the center of Redjacket, where they would march to the Italian Hall on 7th Street. In the hazy light of her plainly furnished bedroom, Annie impatiently made ready for her role. She smoothed the waistline of her simple white dress and was pleased it had a full skirt. Walking would be more comfortable. She combed her heavy hair, and stooping to see her reflection in the aging mirror, she braided, then twisted and pinned the plaits around the top of her head. Her arm muscles, Annie noted wryly, were starting to become accustomed to the stout flagpole and ached a little less today.

She draped the long red and blue ribbons over her arm. "Hey, Joe," she called. "I'm ready if you are."

*   *   *

At the designated assembly-spot, Annie and Joe moved toward their friends who stood in clusters.

"Oh, Annie," one woman said. "You look wonderful."

Annie made a small mock curtsey in thanks. "I thought the streamers might be a nice Sunday touch," she said looking around the congregation, all of whom were especially decked out with white shirts and bows.

"Oh, there they are." She started toward two small girls, each wearing made-at-home white dresses.

"I've tied streamers to the flagpole," she said, bending to the little helpers. "All you have to do is hold on to the end of the ribbon and follow along behind me."

"You won't walk too fast, will you?" the little girl fretted.

Annie laughed, and playfully kicked out one long leg. "I promise."

* * *

As the strike dragged on, the special Sunday parade started either at the town union hall, or the Palestra in Laurium, north a short distance. There, men would address the strikers in all languages of the immigrants.

The strikers, this first Sunday Parade Day, could not help but observe that the militia had arrived and was setting up camp on company property and in other hastily provided shelters in Redjacket. Even more disheartening was the discovery that Sheriff Cruse, *two weeks before the walk-out*, had made provisional arrangements with the Waddell-Mahon Corporation to hire several hundred armed gunmen as strikebreakers. Organized labor knew this company well. Its payroll became active in times such as these. Thugs and parolees lined up by the score at the hiring office.

Some miners, unable to resist the money, also took jobs as strikebreakers. Annie could not tolerate these men.

"It's one thing to be a miner and not want to be part of the strike," Annie complained. "I understand Dad leaving to be with Mary and her family in Minnesota. But to hire out against your fellow workers is rotten and disloyal." Later, when Annie was to find that the company had hired strikebreakers from the Asher Agency, also from the East, she could hardly contain her anger.

There were to be precious few peaceful marches. To the quick- tempered among the strikers, the abuses being heaped on them by the presence of the agencies, the militia and the sheriff's deputies were more than they could bear. "We rightly protest the slave conditions that go with our jobs and they bring in outsiders to beat us down."

Trammers, especially, felt exploited. They were the workers whose place in the mine was to push railroad cars loaded with copper ore. They were on the lowest pay scale for this beast-of- burden job that broke the health of men in seven years or less.

Later, a trammer testified, " . . . we were being forced to do more and more each day of the year . . . it was so bad we feared the company, who made us do the work of mules, would start driving us about like mules."

Another, a driller, said, "Of course, it ain't right—especially when you know that the western miners make more money, with fewer hours than us. And Michigan copper brings more profit to the company than any mines in the world."

Even those strikers with cooler heads were aroused. The union began planning immediate strategies.

"We must hit the pumpers . . . that way we can force the mines to shut down for good until our demands are met." Pumpers were workmen who manned the suction devices that kept underground waters from flooding the mine tunnels. The water was forced to the surface and piped to areas away from the mines.

Accordingly, a band of strikers armed themselves with clubs and rocks and attacked the unwary pumpers when they resisted. The move was successful only until early August, when the pumpers appeared at their stations, bandaged and splinted and accompanied by armed militiamen. The mines reopened, and as soon as the seeping waters had been drawn out, drilling began at about half production.

"It's too late now for anything but a mass protest," the union leaders told their followers. "We must continue to refuse to work until management will listen to our complaints."

But the wealthy, powerful Copper Kings had other ideas.

CHAPTER 7

THE WOMEN RISE UP

The northern Michigan copper mining frontier became, overnight, the hottest news in the nation. Journalists from every large newspaper stepped off the train those first few weeks in July and August. Exhausted from the interminable ride north through the wilderness, their natty suits greasy with soot, they dashed for hotels and rooming houses and a proper meal.

Annie's quick brown eyes missed none of the immigration of reporters and she took every opportunity to present the miner's side. Newspapermen were enchanted with her spirit and forthright manner of speaking. Even editors who were sympathetic to the mine operators grudgingly gave Annie due credit for her verve and conviction. One such writer said the union ought to dress Annie in polished armor and put her astride a white horse with her flag. He dubbed her the miners' Joan of Arc. She was indeed, a handsome woman, slender, shapely, straight-shouldered, and most of the time, poised. She knew the issues, and got her points across in strong, simple language.

"I was born and raised inside the circle of the C. & H. Mines. But these injustices started long before I came into the world. The stinking company carries on a system of paternalism that is degrading. Are we supposed to kneel and kiss MacNaughton's hand because the company builds our homes and schools, then rents them back to us, or donates land and money for our churches? Are we dolts, unable to think for ourselves because we speak a tongue other than English?

No," she declared, "My people have worked since the beginning of these mines, underpaid, in dangerous and unsanitary conditions. Read the newspaper for a week and see how many miners are injured or killed. And we are so unimportant, the editors don't even bother to find out the names of these poor men." Her voice rose, high, mocking. "'A Finn was killed by falling rock,' they say, or a 'Croation's legs were severed when trapped between ore cars.' And now, the greedy company has replaced the two-man drill, increasing the dangers. A man is without help of a partner in case of an accident . . . And God knows how many accidents there are!"

Reporters interviewing the tall, balding James MacNaughton were treated to a story so different, one journalist remarked to another than it was as if they were not discussing the same situation.

MacNaughton would not hear or read the unions grievances.

"I don't consider the majority of the miners to blame," he said. "The whole uprising is in the hands of foreign agitators, and a few gullible men are influencing my workers. I shall continue to refuse to confer with union officials and their lawyer Mr. Darrow. I shall not go to Lansing or anywhere else to negotiate."

In August, the *Detroit News* reported that when MacNaughton was asked how the company would handle non-payment of rent on houses and land, he said: " . . . no advance rents will be collected . . . they can live in them whether or not they are strikers, and there will be no evictions during the time of the strike." He went on, "Also, the bookkeeper will not be deducting the usual dollar a month for the company's medical facilities. These will be at the disposal of all strikers and their families."

Eventually, strikers were evicted from company houses.

*   *   *

Annie did not always march at the head of parades; sometimes a union official or visiting sympathizer carried the leading flag.

MacNaughton stood at the door of his office. Hands in pockets, he watched reporters leave his building and disperse. He knew without reflection it had been two weeks and a day since the work stoppage. He frowned, and lowered himself to the swivel chair behind the desk. His high stiff collar felt too tight.

*　　*　　*

When the *Detroit News*, on August 9, ran its daily report on the Copper Strike, the editors had no inkling of the hatred and bitterness the story was to unleash within the souls of the strikers. A company official (not MacNaughton) was quoted as saying that the C. & H. didn't believe in strikebreakers.

"Either we win or the strikers will win and our old men will go back when it's over," MacNaughton told reporters.

Whether or not the men believed what they heard, some three thousand of them, unable to face further sacrifices, returned to work. Annie was furious. She knew there were strikebreakers hopping off the trains every day, adding to the number of non- union men the company had bribed.

"On top of everything else, they are liars!" Annie cried. She went on leading, in spirited determination, the dwindling parade marchers, through the summertime sultriness. One day the temperature reached 109 degrees.

As the heat of late August made it easier for strikers to leave their beds before dawn to march, women generally began to take an active role in the strike.

The leather of their shoes grew thin and finally gave way to holes. Annie's one pair of black shoes was no exception. She cut cardboard insoles, slipping them inside after each day's long walk.

"I hope these clodhoppers last until we win this fight," she laughed. "There's sure no money to buy new ones."

Miners' women found that Ben Goggia, an Italian immigrant and organizer from the union, struck a strong responsive chord within them.

His job was to put together the daily parades, give speeches and see that they were translated into other languages, and published in the Miner's Bulletin. Evidently his speeches were a tremendous success, for the affected women, very much a part of a miner's life, realized that they were being given an opportunity to protest. That they could cry out and act against the ever-present fear of losing a husband or son to death or crippling injury. For the first time in their hard, drab lives, they could make a stand against the degradation of having to hire themselves out as maids, washerwomen and cooks to the monied class. A people whom they saw as taking more than their share from the copper mines—riches brought up and processed by the laborers, doomed to poverty as long as they worked as miners.

So the women rose up, knowing that the troops and strikebreakers found them difficult to deal with. Women gathered to form lines before the sun came up, and any worker who went back to the mines was fair game. The women cursed, hurled tin cans and eggs, and struck out with their fists and rocks.

The least the scabs (workers who crossed 'no work' lines) could expect from their skirted tormentors were curses and the loss of the contents of their pails. Stray cats and dogs ate well for awhile. The worst that might happen was being whacked with brooms dipped in privy holes! Unarmed, the women even attacked the militia. The army, to their credit, used enormous restraint when subduing them.

Once, when two women were arrested, twenty-five other wives and one man tried to break into the Calumet Armory where the pair had been jailed. Only when the rescuers came face to face with the soldiers' rifles did they turn back. Indeed, Annie went too far. She and a friend, Maggie Agarto, spotted a worker hurrying down 5th Street with a tin lunch pail under his arm. They attacked the unsuspecting man, yelling "Scab!" at him, sent his lunch bucket flying; slapped and bruised him. (The man turned out not to be a scab; he was going to his job as a construction laborer.) Close-by deputies pulled Annie and Maggie off the man, but not before four other women had thrown themselves into the scuffle. All six were arrested and taken to the Calumet City Jail. In a short time,they were ushered to Judge Fisher's Court.

Quickly, a crowd of several hundred strikers converged to find that the attackers had been released on their own recognizance and were ordered to return in eight days for sentencing.

Annie appeared exuberant as she left the jail doorstep, clapping her hands and shouting to the crowd. Secretly, she felt sheepish at the mistaken identity, and promised herself she would stalk her quarry more carefully in the future. But the crowd doted upon her, shouting encouragement, and loving the excitement. Annie waved and had turned to speak to her friend, when a young woman pushed out of the mob and grasped Maggie's arm. Annie knew her as the wife of a Hungarian miner.

"Maggie," the woman cried, "your husband has been arrested, along with Ben Goggia."

Maggie's smile faded. "What has he done?"

"I don't know . . . Maggie, it's probably no more than disorderly conduct. Wait here for news."

Annie remained by the jail with her friends until word came that the two men had been required to post bond, but had already been released.

"Go on home, Maggie," Annie said, patting her friend's shoulder. "I'll see you tomorrow."

*　　*　　*

Annie strode home from the Union Hall to finish her housework and start the noon meal. She was in a dither of excitement. She had heard from union headquarters that Mother Jones would be in Calumet on August 5th.

Dreamlike, Annie worked at her chores, turning over in her mind what she knew about her heroine, whose given name was Mary Harris, but whom everyone now called Mother Jones. Labor papers made much of her fiery agitations for human rights, for "Mother," now eighty three years old, took miner's problems to heart.

45

Annie had read everything she could find about "The Miner's Angel," as Mother Jones was described by a later biographer. Annie knew the spirited Irish-born woman had immigrated to Canada with her parents as a young child, and was raised there. In 1861 she had married an iron worker, George Jones, and had borne four babies. In 1867, she lost her husband and all her children to the dreaded Yellow Fever epidemic which raged through Tennessee where they lived.

By 1871, Mother Jones had pulled her life together. She opened a dressmaking shop in Chicago, but again tragedy struck. The Great Chicago Fire took her home and shop, leaving her nothing but her life.

For Mary Harris Jones, that was enough at first. But it soon became apparent she would have to ask for help. The Knights of Columbus, a fraternal order of her church, came to her aid. Mary became interested, then very active in the organization's campaign for better labor working conditions.

Annie paused in her musings. By the 1880's, when I was born, she thought, Mother Jones had raised herself to a position of importance in the labor movement. Annie had read some of her speeches, and had heard that the little woman was a stirring speaker, never tiring in her calling to the cause of the United Mine Workers.

God knows the coal miners' needs aren't any more pressing than the copper miners', Annie thought. Our men have to go down deeper . . . so far down they have to strip to no more than pants and boots because of the thermal heat, even in winter.

Maybe I could travel around the country, too, for the Western Federation of Miners. It shouldn't be too hard to write a speech, Annie contemplated, as she diced carrots and turnips to make meat pie. Mother Jones' slogan was short and to the point. "Join the Union, Boys." Annie thought about slogans the copper miners might use. She slid the pasties into the oven.

*   *   *

The morning of August 5, 1913, five men whose last names were Erickson, Strizich, Mikko, Oppman and Jedda dashed about town putting final touches on their particular part of the coming week's agenda. Each of these men served the union as organizer for his particular language group. The system worked well, and communication between the miners was excellent. Also, language was no barrier between management and miners. C. & H. officials were often criticized for hiring a high percentage of non-English speaking immigrants in order to strengthen the company's role of paternalism. This certainly seemed to be true in the years before the Western Federation of Mines sent men to organize the Copper Country in 1909.

At the dusty train station in Calumet, a sizeable delegation shuffled restlessly. Everyone connected with the Federation was eager to meet the famous spunky old woman who feared nobody. She had served jail sentences in other states for her views, and although she was desperately needed in Colorado for an ongoing miners' management dispute, she had insisted upon re-routing her itinerary to include the Keweenaw Copper Strike.

With a blast from the engine's steam whistle, the train hissed to a standstill, and Mother Jones appeared, surrounded by United Mine Worker Leaders. Federation men stepped forward, introduced themselves and whisked the party toward nearby autos. Mother Jones, however, insisted upon walking the three blocks to the Union Headquarters!

In town, strikers took their places in the parade line and strained for a glimpse of the first autos carrying the guest-of-honor and other visitors. Annie shrugged into the wide leather straps of the banner holder; someone handed her the huge flag which she eased into the pouch. The train whistle shrieked again.

"They're here!" Word passed down the procession and all eyes watched, as the union delegation arrived, not in the autos which had waited for the famous little woman, but walking briskly in the summer heat! Smiling and waving, Mother Jones made her way smartly to the edge of the long marching line.

47

Annie stood transfixed. She was aware of nothing but her heroine—the heavy white hair, tightly pinned and framed by a soft, squarish hat—a high-necked lacy blouse worn under a dark velvet vest and coat—the black shoe-top length skirt.

The miners, proud to be recognized, started up smartly, smiles creasing their worn faces, new hope in their hearts.

Later, when Annie was introduced to the stalwart little woman, Annie smile down at her, and wondered, as she took the gloved hand in hers, if Mother Jones could detect the tremble of excitement in her own. To think she was truly shaking hands with the famous lady she had read about these past years! Annie wished the moment would last forever.

Within six days, Mary Harris Jones was gone. But she had plenty to say to the assembled strikers, their women and children. She called for more peaceful measures.

"Stay away from spirits and guns," she cautioned, *"If you are attacked, use your fists and black his two eyes. Then he can't see to shoot you."* The newspapers, not long after, reported Mother Jones' imprisonment for her part in Colorado demonstrations.

*   *   *

Sometime in late summer, a young man took leave from his job in Chicago. He was Frank Shavs, an editor on the Slav Union Socialist newspaper, "Proltarec." Shavs' appearance on the strike scene created no large ripple. He was just another reporter among many. But Shavs did not intend to remain unnoticed. The cause had become very dear to him; he had pored over every bit of news that came, daily, from Calumet. He knew his boss, Ivan Molek, a seasoned observer and reporter, had received a call from union officials at the copper mines, for someone to come and help plead their cause. Shavs regaled Molek with reasons why he should be the "someone" to answer the call, and against his better judgment, Molek relented.

"You're too unorthodox, Frank, but maybe they can use you up there," Molek said. "So be on your way."

Shavs left immediately, made himself known to Calumet union leaders, and within days after meeting Annie Clemenc, decided that someday she would be his wife. Annie's marital state bothered him not at all. He saw an attractive, independent woman, childless, giving everything of herself for the miner's struggle. He, himself, had never married. Infatuated, he grew more and more bold.

Shav's intentions, of course, were obvious to all who had eyes. Annie, flattered by the attentions of a big city newsman, ignored the rumors which flew about the mine community. She couldn't ignore the facts, however, that she and Joe had grown apart —- Joe could not understand fully, her passion for socialism and labor justice; Annie had trouble comprehending why anyone would not fight for fair wages and safe conditions . . . as well as human dignity. Added to these problems were their deteriorating finances. The weakened marriage began to crumble as early autumn drifted to Michigan's north country.

CHAPTER 8

ARREST!

Chronicles called it "The Seeberville Murders." The tragedy harked back to an argument a company guard had with a striker because the miner was returning home across company property.

Not long afterward, four gunmen and two local deputies set out for this miner's home. Made of rough, unpainted boards, the two-story house sat on a hillside. In the front yard, that fateful day, several of the miner's friends were gathered to play tenpins. A small area had been cleared of rock and leveled. The players, startled at the appearance of the armed intruders, sensibly faded back when the strikebreakers called for the argumentative striker to present himself. He, just as sensibly, remained within the house and the hunters left. But not for long. They returned and fired into the house where the boarders were having dinner. One striker was killed instantly, another expired the next day. Two other men fell with serious bullet wounds, and a baby being held in its mother's arms was burned with gunpowder. For his part in the ambush, a deputy was knocked in the head with a tenpin. During subsequent court trials, two of the gunmen were convicted of manslaughter.

Annie was sick at heart. She knew the young victims, Dlozig and Steven, as well as the wounded, John and Starka.

"The company will pay for these murders," she vowed. Yet, even as she said the words, inner fears nagged. The company was incredibly rich. Management could continue to buy strikebreakers and scabs. The union had nothing but their courage, and loans that were proving to be too little to spread among the remaining hold-out strikers. Families were already feeling the sacrifice in hunger and accumulating bills.

The remaining days in August found the strikers a bit better off, by the appearance of representatives of the Michigan Federation of Labor. Strikers continued to assault others who chose to work, and were being arrested because of the attacks. On the 25th, acting on a tip, militia men found a bomb under the home of an outspoken union leader at the Mohawk location. The following day, someone discovered a bomb at the mine power house in the same community.

Violence begat violence. Strikers stormed deputies and fourteen- year-old Margaret Fazekas was wounded by a bullet. Women strike-supporters organized, and thirsting for action, again had to be repulsed by the militia.

*   *   *

The Western Federation of Miners' president was a man who had pulled himself up the ladder of union leadership in only sixteen years. Charles Moyer was four inches short of six feet tall, dark-skinned, with a straight, prominent nose and square jaw. His young-adult years had been filled with changes . . . he left his Iowa farm, travelled west and worked as a ranch hand in Wyoming. Romantic notions of what a cowboy should be, went with him to Chicago, where he frequented pool halls wearing a gun slung from his belt. He and the weapon quickly got into trouble.

Moyer and a friend ended up in prison for burglary and strong-arm robbery. When he got out a year later, apparently chastened, he headed for the Mid-west again, married a country girl and took up farming.

The beginning labor movement was the exact vehicle for Moyer's aggressive personality and organizational skills. The Western Federation of Mines had been staggered by its own internal bickering in the 1903-04 Colorado

51

Miner's Strike and was still only partially recovered. Here was evidence, Moyer believed, to regain credibility for the union while getting a handhold on labor reform.

Moyer reached Michigan a week later than he planned. He had been in Austria attending an international Mining Congress when he received a cable saying that the copper strike was underway. He left immediately, and had journeyed as far as London when he was struck by an asthma attack. The ocean voyage aided his convalescence, but the huge task of money-raising for the Michigan workers waited for him in Denver.

On August 31st, when he stepped off the train in Calumet, he had only thirty-six thousand dollars banked to last the strikers until September 12. When pressed by reporters, he stated that the W.F.M. had great amounts of money put aside for the strike.

On that miserably hot afternoon in the Palestra, when he addressed almost three thousand persons (eight hundred children and their mothers) he assured them that the W.F.M. had a hundred- sixty-one-thousand dollars in cash ready for their use. What he did not say, was that this figure was based on loans he hoped to finalize, plus money from a two-dollar assessment on each outstate W.F.M.miner, which was to begin in September. But, he doubled the actual union membership count!

He went on to urge their continuance of morning picketing. "You have the right to try to peacefully persuade others not to work." The governor, he said, should order the militia to disarm the thugs and gunmen and ship them out of the community.

On September 3, he was off to Lansing with the famous lawyer, Clarence Darrow, for a meeting with Governor Ferris. The governor took the crisis to heart. He called for a special legislative conference to deliberate reforms. But it meant extra work, extra time in meetings for legislators. They rejected Ferris' proposal and he gave up. Ferris himself said he believed James MacNaughton, when the mine president said he would let grass grow in the streets of Calumet before he would meet with the W.F.M. or any of its representatives. Labor reform certainly was not getting support from either politicians or management or even some local church leaders.

To Governor Ferris' credit he did not give up on other avenues of arbitration. He personally wrote MacNaughton, with a proposal to leave the W.F.M. out of the meeting of him and his workers. MacNaughton refused.

Indian Summer came and went. Fortunately, the money Moyer promised union strikers became available. Families could survive, at least until the terrible cold season arrived.

On the last day in September, Moyer and Annie stood near the Redjacket location on Seventh Street. Strikers had discovered it was important to be seen and heard, and Annie's tall slender figure and strong voice when she wasn't marching, was very much in the public eyes and ears. Today, she and the union chief were with other strikers, gathered at the curbside, hoping to talk passing workman into staying away from the mines.

"It's hard to keep my hands off the scabs," Annie said, more to herself than her companion. She drew the collar of her long, dark woolen coat closer around her throat. As the sun rose higher, more and more non-striking workers, lunch buckets in hand, appeared.

Annie called to a miner as he approached.

"Where are you going, partner?"

The man slowed. *"To work."*

Annie stifled the impulse to scold, and softening her attitude, said, "Not in the mine, are you?"

"You bet I am!"

Annie began her sales talk, by now well polished, and the miner flattered by her attention, chatted amiably. Presently the miner's wife stepped forward; touched his arm. He tipped his hat to Annie, and eyes downcast, hurried toward the shaft entrance. Annie didn't try to keep track of her unsuccessful attempts, but she felt a childlike delight when she was able to convince a man not to cross the picket lines.

"Annie, with the C. & H. president Shaw refusing to leave his comforts in Boston and MacNaughton turning his

53

back on arbitration," Moyer said, "things are bound to get worse before they get better. We must be prepared for that."

"Yes, I know," Annie said, her forehead wrinkling. "I have no children, yet we're finding it harder each day to make the food stretch. If it wasn't for the few dollars I have coming in from my boarders, I don't know what we'd do. Even families who have gardens are running low because they share with the rest of us."

Moyer shuffled his feet and looked around at the barren trees and gray skies. "Don't you usually have snow by this time?"

"Sure do," Annie said, "We're lucky so far. At least the streets are clear for marching."

Presently two Slavs, one of whom Annie knew, approached. Running to face the pair, she cried, *"Oh, George! You are not going to work, are you! Don't allow that bad woman (his wife) to drive you to work. Stick with us and we will stick with you!"*

George hesitated. His face flushed and he ran a rough hand through his blond hair. Embarrassment changed to determination and he left his friend to step out of the stream of workers.

Strikebreakers, never far from any public appearance by the union members, materialized. Two of them grabbed George by the shoulders and propelled him along the street.

"You coward," they accused, *"Are you going back because a woman told you not to go to work?"*

George, not so much angered at being called a dastard as he was resentful at being pushed around, dropped his precious lunch bucket and fought back. Immediately, six more deputies jumped into the fracas and George found himself being dragged toward the mine.

The swarm, which was beginning to look like ants taking a prize bit of food to a hole in the ground, paid no heed to the defiant Annie who shouted and gestured after them. Her church training notwithstanding, Annie was familiar with the coarse language of the laborers, and knew how to use it. She was in good form that morning.

"Annie, you have to get away from here."

The man's voice sounded from behind her. It was clear, unemotional, commanding. Startled, Annie wheeled to face an army officer, a stranger.

"No, I'm not going. I have a right to stand here and quietly ask the scabs not to go to work."

The officer politely ignored the fact that she had been anything but quiet.

"You will have to get into the auto," the officer said. Annie drew herself up, and with crossed arms assessed her position. Somewhere a child laughed, a mine whistle blew. She smelled the oily exhaust of a chugging automobile engine; saw the officer's aides sitting inside.

"I won't go until you tell me the reason," she said. Annie had lived these past weeks under the constant threat of arrest; she felt no alarm. Without warning, two of the commander's aides appeared on either side of her, and half-pulling, half carrying, hoisted Annie on to the seat of the first automobile she had ever been in.

Outraged, Annie pounded her feet on the car's floor-boards. Powerless as her position was, she continued to stamp her feet and demand an explanation. But the aides only climbed in the vehicle; slammed the heavy, black half-doors. At a signal from the ranking officer, the driver turned the wooden steering wheel toward the city jail.

Annie was soon to learn that her abductor was General Peter Abbey, commander of the state militia.

Bumping along Calumet's streets, the general leaned back and looked squarely at Annie. *"Why don't you stay at home?"*

Annie stopped banging her heels and returned his stare.

"I won't stay at home," she said. *"My work is here. Nobody can stop me. I'm going to keep at it until this strike is won for the workers."*

The car trembled to a stop and the aides escorted Annie to the *"dirty little Calumet jail,"* words she used later to describe her experiences. She was charged with assault and battery.

"I haven't assaulted or battered anybody!" Annie protested to the jail clerk. "These men assaulted me! I was standing quietly rallying the men to hang together. I wasn't breaking any law!"

The clerk looked over his glasses at Annie and the soldiers standing by.

"I wondered how long it would be before you'd land in here, again," he said, not unkindly.

*   *   *

Eight days before Annie's arrest, county Judge Patrick O'Brian, although sympathetic to the union cause, granted an injunction instigated by the mine's attorneys to stop all parading and picketing. The union, naturally, chose to disregard the order while their lawyer, Clarence Darrow, swiftly drew up papers pointing out that it violated their constitutional rights. A week later, on the 29th of September, the judge negated the order, on the grounds that it should not have been granted in the first place.

*   *   *

At 12:30 in the afternoon, after five hours of imprisonment, Annie walked out of the jail building, head held high. She went directly home where she wrote an article telling of her seizure and confinement. It appeared two days later in the *Miner's Bulletin*, a union newspaper, which ran it on the front page. It was entitled "A Woman's Story" and had no byline . . . possibly because the editor felt that none was necessary. Not only had a sizeable portion of the community witnessed the incident, the remaining citizens had heard it from word-of-mouth.

The general had been busy. Within hours of Annie's arrest, he gave orders to halt the morning parades, his motives being, reasonably enough, that crowds were potentials for violence.

The unsuspecting marchers and their families were unprepared, then, that first day in October, when mounted militiamen began riding into their midst shouting, "Disperse! Go home and eat breakfast! No more public demonstrations!"

Annie sidestepped, avoiding collision with a horse, but refused to leave the street.

"You can't prohibit a peaceful march," she cried after the cavalry. "Go back and tell that ninny Abbott that we will march! You'll have to kill all of us to make us stop!"

At that moment, a solder's horse turned toward a group of marchers who had fled to the sidewalk. Annie felt, rather than heard the screams, and when the soldier had backed off his horse and the fallen had picked themselves up, a little girl lay motionless, her head crushed by a hoof. Her mother knelt, mute with shock, and began frantically to mop up the heavy welling of blood with her shawl.

"My God!" Annie cried, thrusting the flagstaff into the arms of a striker nearby. "What have they done!"

She picked up her skirts and sprinted to the scene. Someone had bound the child's head with a handkerchief which was stained crimson.

Willing arms carried the limp body to the company hospital. The doctor shook his head. "She's very bad off," he said. "Although there is a small chance she will live. Even if she does, there may be extensive brain damage."

(But the young girl, after hovering for days at the tunnel of death, regained consciousness and surprised everyone by recovering completely.)

The following day, still hot with indignation at the treatment she had received at the hands of the militia, Annie stood with strikers at the South Hecla Mine. Her anger grew as she noted more and more men, strangers, being imported to replace the dwindling ranks of strikers.

"Hey, there, Scab!" she called out to a workman passing. He was nearly to the mine shaft. "Don't go down there. It's blood money you're making."

The miner turned away from her, and Annie, infuriated at his rejection, spat in his face. Militiaman Major Britton witnessed the insult and in two long steps was beside Annie, restraining her arms. Immediately, a striker fell upon the major and began to choke him.

"Let her go or I'll squeeze your life out," the miner cried. Major Britton, his face fast becoming a liverish color, released Annie, struck out at his attacker and freed himself.

From there the women and deputies went at each other with fingernails, fists, clubs and rocks.

CHAPTER 9

BREAD AND ROSES

Annie and nine other women were arrested and, again, let out on bond. One reason copper miners' women became active could well have been the Lawrence, Massachusetts, textile workers. These laborers were women; girls, really, who were being abused almost as badly as their male counterparts in mines. The textile workers organized parades and marches, and in one of them, a young woman carried a banner reading, "We want bread and roses, too."

James Oppenheim, a poet, was touched by this gesture, and penned four verses of such stirring lyrics that Caroline Kohlsaat wrote music for the poem. The song became tremendously popular all around the country, and the textile union was quick to use it as a statement of workers' demand for justice.

The last verse, particularly, would have special meaning for the Copper Country women. To them, roses were their men's need for independence and dignity; the bread was decent wages for the miners, so their women would not be forced into demeaning jobs.

> As we come marching, marching, we bring the greater days.
> The rising of the women means the rising of the race.
> No more the drudge and idler—ten that toil where one reposes,
> But a sharing of life's glories: Bread and roses! Bread and Roses!

*　　*　　*

Ben was an Italian immigrant who had risen to importance as a union organizer. For three dollars a day, he took responsibility for forming the daily parades. He gave speeches at meetings (in both Italian and English) and saw that his words were printed in the *Miner's Bulletin*. Ben was forceful, charming and articulate.

So the miners' women, for the first time in their hard, drab lives, were being given the opportunity to lash out against the poverty and ever-present fear of losing their husbands, brothers or sons to death or crippling injury. Especially unjust was the degradation of having to hire themselves as maids, washerwomen and cooks to the moneyed class, people whom they saw as grabbing more than their share from copper that was brought up and processed by the underprivileged.

The women rose up, and knowing that strikebreakers and troops found them harder to deal with, gathered and formed lines near the mines in the very early morning. They fought strikebreakers, scabs and militia any way they figured would be effective, lawful or not. They went after working miners with filthy brooms; they rampaged with rocks, sticks, fists and spit. Ben did not encourage the women toward lawlessness, neither did he discourage their zeal. Union leaders knew well that women were crucial to the morale of the strikers.

The children, even, were given an active role. The first week of October, a plan for a children's protest was hatched. The date was set for October 6 . . . the following Monday. Instead of reporting to their classrooms, the strikers children outfitted themselves in their "good" clothes. Downtown, older children stood tall and proud beside the young ones, who were too excited to stand still, let alone look exalted. They formed two rows and led the parade of adults. A 'big boy' wearing knee pants and a long-sleeved white shirt, gravely carried a huge, hand-lettered sign that read, "Papa is Striking For Us."

60

The students were delighted at being part of the protest and many begged to do it again.

"You need to get an education," their parents said, "so when you grow up you won't have to do what we are doing."

Meanwhile, downstate at the Capitol building in Lansing, the Supreme Court overturned Judge O'Brian's dismissal of the injunction. By October the strikers found themselves being forced off the streets by armed soldiers and guards. But day after day the fighting continued, each side shooting, dynamiting, ambushing the other. Striking miners made no progress. Mine operators, while not prospering as greatly as before, had convinced scores of workers to return. Many workers themselves, unable to bear the sacrifice, gladly went back. Some twelve hundred employees hired from outside the community brought up the copper that was being loaded into railroad cars, pulled to Houghton-Hancock and shipped to big cities in the East.

With November, came icy gale-force winds, but except for a blizzard and thaw on November 8, no snow fell. It was a peculiarity which was to remain through until January. On this day, though, ninety-nine strikers were arrested, Annie included. She had taken up her flag and slogged, facing the wind and sleet, south down Calumet Avenue. There the strikers accosted mine workers, but were taken into custody before they could use the razors, brass knuckles, blackjacks and guns concealed in their clothing. They were desperate people using desperate means. The C. & H. Company had reneged on their promises. Families who couldn't pay the rent were evicted and the hospital was closed to them. The biggest lie of all was MacNaughton's vigorous denial that the company hired strikebreakers. He told a Detroit reporter, *"I have no use for Waddell (the strikebreaking company) or his class . . . we have troubles enough without having that kind of men brought here."*

The incredulous strikers went on to read, *"I would just as soon have 'Lefty Louie' or 'Gyp the Blood' in our employ as a peace officer as James Waddell."*

"Who does he think he's fooling," Annie exploded

61

when the general manager's words were repeated at a union meeting. "Those thugs have been here since the beginning. Everyone is terrified, not just the strikers. See how everyone in town keeps their window shades pulled! And it's common knowledge that MacNaughton is the boss of the County Board of Supervisors. Nearly all the men on that commission furnish goods or services to the mines and they eat out of company hands. MacNaughton tells them when to stand and when to squat. MacNaughton not only gave his approval for hiring strikebreakers, you can bet he instigated the move."

During the month, more than 70 strikers were arrested for violating the restraining order. They picketed and paraded with a vengeance. Once, receiving a report that the railroad was carrying strikebreakers to Calumet, they gathered south of town and stoned the train. This began a series of terrorist acts that killed seven strikers and injured another child. Annie was greatly saddened, but not surprised. She was to learn, some months later, that a hired gunman confessed that his employer (Waddell-Mahon) had given instructions to make some "business."

"Get out there and stir things up, men. It will keep you in a job, you know."

What the strikebreakers didn't know was that the agency was profiting at the rate of two dollars a day on each of them. This had to mean that management was paying out more than twice as much for intervention as they paid the miner, who broke his back a mile underground for twelve hours a shift. The one optimistic happening in November was the appearance of Federation stock stores. Merchandise was limited in variety, but credit was assured.

Nothing progressed in the way of a settlement. The mining company's management refused to believe the miners had legitimate complaints, contending that the "Foreign Red Socialists" were at the bottom of all the trouble. True, the Socialist Party had written kind words about the miner's plight and had sent representatives to Calumet as supporters. The poor strikers needed all the aid they could muster. In fact, the Miner's magazine reported that the counsel for

C. & H., Allan Rees, asked Clarence Darrow to "advise the leaders and organizers of the Federation, not only to waive recognition, but to withdraw from this district and give up their attempts to organize the same; there would be no trouble whatever in settling this strike and no arbitration would be necessary, for the reason that in such a case we would have no difficulty in coming to an entirely amicable arrangement with our men."

The owners and managers, themselves would not even consider an arbitrator, such as the President of the United States.

James MacNaughton said emphatically, *"This is my pocketbook. I won't arbitrate with you as to whose it is. It is mine. Now it would be foolish to arbitrate that question . . . "*

The many weeks since July 23 had seen men from the United States Department of Labor, Governor Ferris' staff, and Congressmen try to arbitrate. No one could persuade the companies to bargain.

Each side, of course, had a right to feel outraged. The miners still hanging on, a pitiful few, considering the thousands that packed up and left the area, clung to the hope that their suffering would make changes, that their children and grandchildren could remain in the place of their birth and earn a safe, decent living from the mines. Unfair wages, interminable working hours; certainly these were basis for protest. Yet the most deeply felt need was recourse for complaints. Men could go no further than the mine captain, the god of the depths. If the captain chose to cheat (he did all the rock measuring for which the miner was paid) or if a worker had a disagreement with his captain, there was no recourse. Traditionally, mine managers set aside a day and a half each week to hear individual workers problems. Hat in hand, the miner was expected to accept without doubt the administrator's assessment. The practice was seldom used and helped no one except perhaps the officers' egos.

Sadly, grievance procedures were not good public relations. People understood hours, wages and hazards. Communication and human dignity would not have presented a strong case for the laborer. The Federation, therefore, down-played this aspect, and the result was a misunderstanding on the part of observers willing to accept the side of the company mines. It seemed as if management was ignoring everything except the "threat of Socialists and Reds and Communists."

As Lake Superior winds grew colder over the narrow peninsula, so did the hearts of the strikers. Some, pushed to the hard edge of hatred, ambushed a company rental house where three non-union Cornishmen slept.

The young men had just returned to their mining jobs, being told by their friends that conditions had stabilized. They died, and others in the house were wounded. Sadly, no one was found and charged for the murders, and feeling ran high against the union. So strong was public opinion that businessmen and community leaders started up an "Alliance" calling for the union and its "Red" supporters to leave town. Merchants began cutting off credit. No more purchases "on tick."

*   *   *

With apprehension, Annie boarded the electric streetcar to Houghton, seat of the county government, 13 miles south. As the car clattered and swayed over the gray landscape, Annie thought about her estrangement from Joseph and friends who were now dead because of the strike. She was troubled by these events. The only positive thing in her life seemed to be the deepening relationship with Frank Shavs.

At the courthouse, Annie was given a jury trial, and with no lost time, found guilty of two counts of assault and battery.

"Isn't it interesting," she said to Shavs after returning to Redjacket. "Nobody on the jury had a foreign-sounding name. Well, I have to go back and face the court again . . . the judge has delayed sentence until January. I suppose I shouldn't fret," she said, smiling into Frank's dark eyes. "I'm free to help with the effort."

64

"Yes," Frank said, taking her hand. "You're a symbol to the strikers —seeing you at the head of parades with your flag means the fight for reform is alive."

The Citizens' Alliance, birthed by the Calumet and Houghton merchants, grew quickly. Well fed by experienced parents, the organization drew others to its family where they actively attacked the problem of dwindling purchasing power. The first weeks of December had come and gone, and the usual pre-Christmas surge in sales didn't happen. The Alliance had a lot to say about the murders of the Cornishmen and nothing at all to say about killings suffered by the strikers. Citizens' Alliance also paraded and purchased quarter-sized metal buttons with their name printed on a white background. Members were admonished to wear these buttons whenever in public.

At this point, the strikers' ranks had fallen off until there were but a few hundred holding firm. But they drew together; talk at their meetings tempered. Annie had to accept the gloomy picture of company gunmen (goons, they were to be called later), the county's strikebreakers and now the condemnation of their friends and neighbors in the Citizens' Alliance. Strikers concluded that they must slow down to survive.

Another new-formed group was the "Women's Alliance" but it was non-political. With other miners' women, Annie met and organized the group for one purpose only— to make Christmas a little jollier for the miners' children. Many of these unfortunates came home from school to a cold house and cornmeal mush for dinner—with luck there would be brown sugar to stir in. This Alliance sent passionate pleas for contributions to unions in other cities, and as the Christmas season edged closer, the women were heartened. Each day the train unloaded boxes of candy, clothing and gifts from distant places, which the striking miners, glad for the chance to relieve the boredom of unemployment, tenderly transferred to horse-pulled wagons. Singing Christmas carols in a half-dozen languages, the volunteers stomped up the stairwell to the stage-auditorium of the Italian Hall with the precious cargo. There, the women grouped the contributions according to kind, and on long tables set up on the dance floor, the wives, sisters, mothers, older children and grandmas sorted, bagged and wrapped. It was warm, cheery. All wanted to forget the presence of the white, wooden crosses nailed to telephone poles wherever a Waddell-Mahon hired hand had killed a man. They were weary and ashamed of the terrorism done by their own as well, and with the word out that the company was promising an eight-hour day for underground workers and nine hours for surface men, they believed there was progress.

"If we go to work Saturday night, instead of to the pub, we get ten cents added to our paycheck," some of the better-known drinkers scoffed. Yet some family men did choose to go underground the last shift of the week to earn an extra ten cents.

As the days grew shorter, Frank Shavs' admiration for Annie became more open. Strike Committee officials complained to Shavs' Chicago employer that the reporter's presence in Calumet had already done more harm than good. The Slav-Croatian community was upset because many newspapers described Annie as being Croatian. "Our Croatian women do not act in such a manner," was the gist of letters that protested both the inaccuracy and Annie's behavior.

To the union's grumblings, Molek replied that they shouldn't try to *make a horse out of a mosquito*" and Shavs did remain in Calumet until he felt like returning to Illinois.

As for Annie, she had enough loyal followers, some of whom could not have been adverse to a little romance amid the hardships. Whatever Annie's feelings were early in her relationship with Shavs, she held them privately.

December dragged on, but Charles Moyer remained cheerful.

"Look, friends," he said, "Incidents are fewer and the companies did make concessions to the men who went back to work. Granted, it's not what W.M.F. wants, or what you want, but it is progress. Have faith. You may not have to sacrifice much longer."

66

Secretly, Moyer worried about an ultimatum the companies had thrown at strikers—return to their jobs by December 19 or be permanently replaced by imported workers. It was a blow that hurt. With their jobs taken by others, strikers would be forced to leave the district and seek employment elsewhere. Adding to that concern was a ruling by the Michigan Supreme Court making permanent the injunction against picketing and parading. Annie, Moyer and other union officials who planned strategies, stewed about this restriction, for the militia was rigidly enforcing the law.

Lastly, a grand jury had been formed to investigate terrorism.

Nobody had any illusions about the bad public relations and press coverage Western Federation of Miners had received. The hometown paper, *The Calumet Miners' Gazette* was hostile and greatly damaging, as much by what it chose not to print as by what it did print. Many times the lawless acts of goons Waddell-Mahon hired went unreported. The newspaper also ignored the visit of John L. Lewis, who later rose to be champion of coal miners and a world famous figure.

Annie always believed the strikers would get their demands although she was not naive enough to believe that it would be fast or easy. Mother Nature cooperated, however, and the mercifully snowless December days allowed smooth marching, in spite of the injunction. One day the miners strutted down the main street, each wearing a Citizens' Alliance button pinned to the seat of his pants. Three deputies were fired on and a near riot ensued. So much for the 'slow-down' policy of the strikers and the sense of humor of the other side.

CHAPTER 10

ANNIE MEETS ELLA

Ella Reeve Bloor walked into Annie's life a few days before Christmas. Then fifty-one, Mrs. Bloor was a woman of strong convictions. Hers was a firm, handsome face with dark, expressive eyes; she had curly brown hair which she wore parted in the middle and pulled back to a soft bun.

A Socialist Party member, she believed that workers' rights needed to be fought for, and she held a particular interest in the welfare of miners' wives and children. Hearing about the hardships of the copper miners, she took herself and her considerable journalistic talent to Calumet.

From the bitter northern winds of late December, Mrs. Bloor hurried to the blessed warmth of the station-master's office. She took a moment to question the where-abouts of the Women's Auxiliary, and shortly knocked at the door of Union Headquarters of the W.F.M.

Annie, chairing the meeting of the auxiliary, answered the rapping.

"I'm Ella Reeve Bloor from New York, and I've come to see what I can do to help your women and children."

"Oh," Annie said, surprised. She assessed the no-nonsense but attractive appearance of the newcomer. "Step in. Do you have a union card?"

Mrs. Bloor dug into her handbag and produced a Mine Worker's Union Card as well as a Socialist Party Card.

Annie's wariness disappeared and her face lit up. "I have one of those too." She smiled and indicated a large

Socialist button which was fastened to her blouse. "Let me take your coat."

Pleased at the new turn, Annie introduced the guest and resumed the meeting.

"Ladies, we must be absolutely fair in distributing the donations and contributions from the other unions. Check for families with the most desperate needs and take care of them first—each home must receive a proper share."

Next, Annie brought up the entertainment part of the Christmas Program.

"They should not be deprived of their Christmas because of the strike," she said. "I'm trying to collect money for little gifts."

The auxiliary gave Annie the go-ahead and the meeting ended. Afterwards Annie and Ella lingered to talk and began a friendship which was to become very special to both of them.

The following day, Annie took the sixty-four dollars she had solicited from businessmen and boarded the electric train to Houghton and Hancock. Mine-owner agents were always near when Annie was in public and they trailed her all over both towns as she carefully shopped for mittens, stockings, inexpensive toys and candy. Annie paid little attention to the men, not wanting to grant them the satisfaction of knowing they annoyed her.

*   *   *

On December 24, Clarence Darrow received notice that his application for an injunction against the Citizens' Alliance had been granted. There had been more skirmishes.

"You're barking up the wrong tree," Annie pointed out. "Waddell-Mahon thugs are at the bottom of most of the crimes against us."

But Annie was not a W.F.M. officer. Consequently, she held little political power. She was attractive, popular and forceful. She made wonderful news copy.

*   *   *

Annie moved briskly along Pine Street. She was grateful for the snowless sidewalks. It meant she would arrive at the Italian Hall with dry feet. The temperature was well below freezing, but Annie hardly felt the cold, so busy was her mind on Christmas Party plans.

Lifting her skirts, Annie ran up the steps which ended at double doors.

Once inside the hall, she paused a moment. Second floor sounds drifted down—a child pounding on a piano, rapid footsteps, a burst of laughter. When she reached the kitchen, her spirits lifted. Already working were her mother and little Frank. With a bright smile she hugged Mary and leaning down, took her brother's face between her hands and kissed his cheek.

"Hey, your hands are cold!" he said. "And, Annie, we were the first ones here. Mama, when can we eat?" he asked in one breath.

Mrs. Klobuchar tipped the cake-cutting knife at him." "Out! We can't work with you hanging on our aprons."

"Come on, Frankie," Annie said. "You can run down to the basement and see if the men are ready to carry up the barrels of chocolate drops and presents."

Frank galloped away; Annie turned toward the auditorium. Stepping up the stage she pulled a handwritten program from her dress pocket. Frowning, she read:

> Ballerina
> Christmas carols
> (English)
> (other)
> Recitations
> Adult Dancing
> Mother Goose Play
> Santa and Gifts.

A simple evening's entertainment, yet she had spent hours making arrangements. She paused at the spruce tree she had decorated with left-over pink and blue crepe paper and ten cents worth of glittery tinsel. There were no candles; fire was an ever-present spectre in their lives. She ran a hand over one of the boxes containing three hundred Santa stockings the Federation had sent.

She looked up as Frank panted across the big room and leaped onto the stage. "They say they're ready and will be right up."

"Thanks, Frankie," Annie said, patting his shoulder. "It's an hour before the party starts. Try to stay out of trouble, will you?"

Frank bobbed his head. "Maybe I can be first in line for Santa," he said hopefully. On second thought, the room was beginning to buzz with children. He might do better staying close to where the refreshments were being readied.

By two o'clock, the hall bounced with celebrating souls, all bent on forgetting the deprivations of past months. Annie's now practiced eye gauged the crowd to be about two hundred fifty adults and perhaps five hundred children. With a sinking heart, she realized there would not be enough hang-by-the-fireplace stockings. She would have to hold back all the gifts sent by sympathetic unions and use these as replacements for the Santa Stockings.

On stage, Annie fumbled for the opening between the closed curtains and stepped through. The candy and gifts could be stored backstage, she thought, and then set in front of the curtains as needed.

The volunteers staggered up from the basement depths carrying the cartons and barrels. She motioned to them.

"Back here, please." Annie indicated the stage wings, then referred again to her program schedule. The ballerina was first on. Better see if she were costumed and the pianist ready. Annie wanted the entertainment to start on time. The committee had planned the afternoon so that the many unescorted children who had walked, some of them long, chilled miles, would be off the streets before sundown. "On Christmas Eve, children should be home," Annie said, pursing her lips.

*　　*　　*

From the stage wings, Annie watched as the audience sang Christmas Carols, a boy recited a poem and a Finnish

71

man spoke of the Nativity. Mrs. Sizer, close beside Annie, left to seat herself at the piano as applause waned. The audience warmed to their part in the program, and sang from memory, in the language of their birth. The last songs were English and American.

Grown-up dancing came next. Couples circled and stamped, while downstairs, the men in Vairo's Saloon rolled their eyes toward the ceiling and laughed, saying it sounded like the union was trying to dance away its troubles.

*   *   *

Anxiously, Annie checked the sacks of gifts and candy boxes. There seemed to be enough to go around. She heard footsteps behind her, and turning, saw that it was Mrs. Elin Lesh. "I'm worried about the timing, Annie," she said. "Maybe we should postpone the Mother Goose Play until after the gifts are passed out to the youngsters."

Annie sighed and peered at the crowded auditorium.

"You're right, Elin," she said. "We could give out the presents to the children and adults, too, and have the play later in the afternoon.

"Settled!" Elin said. "I'll make the announcement and we'll get started."

*   *   *

At the big doors that opened to the hall on the second floor was stationed a man—a union member whose task was to check the identification of every person entering the immense room. One either had to show a union card or a Women's Auxiliary "book" to be admitted. All was orderly; the Finnlander at one door testified later that the only outsider was a child who attempted to crash the party. He was admitted but not given a gift.

Committee women lined up the children for Santa at one door, directed them across stage where Santa sat; from there they shuffled out of the room from a different door. Annie stood near Santa, in front of the draw curtains, and handed gifts also. Mrs. Therese Sizer remained backstage, handing supplies through the heavy curtains as needed.

Excitement grew; each child wanted to be the next to receive the much anticipated present. The polite line disintegrated. Children surged onto the stage, pushing and jostling one another.

Annie put down the gift she had just picked up and drew herself up to her full height.

"You will not get a present until you sit down," she said loudly. Santa leaned back in his chair and waited also. Reluctantly, the children went back to the rows of seats. They were still excited and noisy but trying. Annie and Santa, seeing some order restored, resumed the gift-giving. In moments, the crowd had forgotten their good intentions.

Disgruntled, Annie stalked off the stage and threaded her way to the man attending the door.

"If they don't sit down," she said, indicating the chaos at the end of the room, *"they won't get anything."*

The man nodded and started forward. Annie hastened back to the noisy stage, not noticing that Mrs. Sizer had left her place to step down on one of the two desks which sat close to the stage. Nor did Annie hear Mrs. Sizer pleading with the children not to rush; that there were enough presents for everyone.

At that moment, a man's voice, clear and powerful, cried "FIRE!"

CHAPTER 11

FIRE!

He repeated the alarm in Croation.

Annie's hands froze holding a gaily tied package. She recovered and shoved it into the arms of the nearest child, and strained to see from where the shout had come. Near the doors, she thought, somewhere by those hallway doors! It was dense with partiers. Children had begun to scream and weep and run about like frightened deer.

A man holding the hand of a child on either side of him, ran to Annie.

"Is it sure there is a fire?"

"No," Annie said, and scurried to the stage curtains. *"There's no sign of smoke or fire. I can't see anything like that."*

"Gee, I'm scared," the man's voice trembled.

"Listen to me," Annie said. *"Stay here with your children. Don't go to the door."*

From the corner of her eye, Annie could see crowds surging for the exit door in the back of the hall. A picture of her mother and little Frank flashed to her mind and she prayed they would not panic.

*   *   *

Theresa Sizer, at the cry of "FIRE!" had jumped down from the desk and shouted, *"There is no fire!"* Some adults had slowed, others stampeded. Sprinting toward the stranger whom she had spotted from her perch on the desk, Mrs. Sizer reached the man.

"What are you saying!" she cried, looking up in his face.

She saw a mustache and dark eyes above robust cheeks. *"There is no fire!"* she repeated. Hardly aware of her actions, she grabbed the heavy, dark overcoat and shook the lapels.

"See! See! Back! Fire!" The stranger said earnestly, waving his hands.

Pushing away from him, Mrs. Sizer dashed back to the stage. The stranger, still yelling, "Fire" ran from the room.

While this was happening, Annie remained on stage, vainly trying to hold down the children's panic. She saw Mrs. Lesh hurry over to speak to Theresa Sizer. Suddenly, the building's fire alarm sounded; its shrill clanging set off the already terror-stricken into mindless crushes at the exit doors.

"What will I do?" Mrs. Sizer looked at Mrs. Lesh, pleading.

"I know!" she said, suddenly resolved. "Hurry . . . raise the curtains!" she called, and four men rushed to tug on the rope pulleys.

They were unable to make the equipment work, so Mrs. Sizer fumbled her way arouhd the draperies and began to play the piano.

Annie paced about the stage. "Children," she said touching them lightly, "Stay here. See—there is no fire. Smell–there is no smoke. Stay with me. Don't go to the doors. We will be all right."

She had to raise her voice over the strains of the piano. Theresa must be trying to drown out everyone's fears, Annie thought. She scanned the crowd—no sign of her mother or her brother. She wondered how long it had been since the ghastly cry of "Fire." A small girl ran kitty-corner from the hallway to the stage and looked at Annie. Her eyes were wide.

"It's full of deputies out there. Something's going on."

Annie left the stage at a dead run, threading around overturned chairs and tables. She had just reached the hallway doors when they burst open and a deputy appeared with a child in his arms. Annie stumbled and halted in her flight, uncomprehending. Spying Annie, the deputy thrust the child at her and Annie instinctively clasped the limp body to her breast. Somewhere she found water to pour over the little, white face. Heart twisting, she knew the child was dead. Dead! And she did not even know whose infant it was.

The hall doors opened again and again. Annie recognized city firemen; they too carried lifeless children. By now, the auditorium had erupted into cries of horror at the awful realization that some tragedy lay behind the double doors. There was a thunderous rush toward the exits; parents, brothers, sisters, grandparents—all praying that their loved ones had been spared; crazed with the need to know; fearing what they might find.

Annie, yet too stunned to grieve, felt her legs take her to the row of little bodies that lay on the floor near the Christmas tree. She knelt and placed the child beside the others. Rising, she forced herself to scan the forms of the victims. None of them was Frankie or Joe. God only knew what was really happening in the hallway. Panic drove her to face a deputy bringing in another. She moved toward him, screaming,

"Are there any more children dead?"

"What's the matter with you!" he answered. *"None of these children are yours, are they?"*

His gruffness released in Annie the anguish she had been repressing. Her conscience shrieked with guilt. This party was your idea, it cried. You, Annie, are responsible for this! These babes are dead because you arranged everything. You even decorated the tree that they lie under, never to see another Christmas. The tears came, then, streaming over the high cheekbones.

"Yes! They are all mine. All my children!"

At that moment, several priests appeared and hastened to the death scene. With a sudden bitterness, Annie let loose the grudges she had been harboring against the church. From the beginning, Catholic clergy had preached against the strike, castigating those who participated in the uprising. To crucify justice in labor because the mine companies bought favor with occasionally donations of land and money to parishes, was an unredeemable sin in Annie's eyes. Shrieking, sobbing, she threw herself at the shocked priests, pummeling them with her fists.

"Don't let those scab priests touch these children."

Annie was quickly pulled away by firemen and deputies. Two of them escorted her, still weeping, to the courthouse where she was locked up.

It was to this building all the bodies were brought later.

*   *   *

At first, kitchen workers, behind closed doors, did not hear the fire cry. They were startled, then, when people came streaming through to the fire escape at the back wall. Hardly anything in the room escaped the crazed human flood. Women and men fainted; a teen-ager cried, "We're being gassed with poison!"

Mrs. Klobuchar pulled Frankie close to her.

"Get over there to the fire escape. I don't think there is really a fire, so don't worry. Just climb down! I'll try to find Joe."

His heart fluttering with fear at leaving his mother, Frankie edged toward the window. Once outside and descending, he clung to the metal railing so tightly his hands ached. At the last rung, he hesitated. Should I jump, he asked himself, looking hard at the gap of several feet to the ground, or let myself down over the last step and drop. He was spared the choice when the body of person directly behind slammed into his back. Frankie catapulted to the frozen earth; winced as pain stabbed an ankle. He lay, enveloped with hurt. Someone on the ladder was bellowing for him to get out of the way. Frank pulled himself to his feet and limped a few steps.

Someone else gently took his hand. "Come on, Son, we'll put you next door for a while."

Frankie found himself, along with other refugee children, huddled together in the parlor of the Van Biber house. Eventually, he strayed outside to the front yard, where he watched, numb with shock, as frantic townspeople carried the lifeless forms from the stairwell. He fretted about his brother. Joe had not wanted to go early to the program. "I'll walk over when it starts," he said. But it was darkening, and Frankie could not tell if his family was among the victims.

He was bewildered, lonely, scared. His ankle throbbed. Maybe, he thought finally, everyone is already home. Or what if nobody is there?

He wished heartily his father had not left when the strike started. They received a letter now and again from sister Mary in Minnesota, saying that they were all well and George had found odd jobs so he wasn't under her feet all day long.

Frankie hobbled home, and with a sinking feeling, saw there were no lights on inside. He let himself in, threw some wood in the stove and settled by the window to watch for his mother and Joe. When they appeared out of the darkness, his joy was boundless.

"I knew you were all right, Frankie, but it took me awhile to find Joe. Thank God you and Annie are all right," his mother cried, hugging him.

It wasn't until he was curled in his bed that night, he realized he had not received his present at the party.

*   *   *

When the terrible news reached MacNaughton's home, he sent company doctors and nurses and ambulances. Mostly, the company medical services were unused—there were only five injuries requiring medical attention. But seventy three persons died, thirty-five of them children.

As Christmas Eve darkness shrouded the stunned community, volunteers carried the remaining corpses in autos to the town hall and placed them on tables for identification. From there, many of the bodies were again wrapped in blankets and taken home for prayers. Only an hour-and-a-half had gone by since the panic began.

The nations' newspapers, many previously unsympathetic to the miners' strike, reported the disaster with compassion. Governor Ferris offered to *"render all assistance possible, in this, their hour of sorrow;"* the dailies quoted him on Christmas Day. Sympathizers sent money and offers of all kinds of assistance after reading the shocking details. The Citizens' Alliance members put their prejudices away and collected twenty-five thousand dollars for the bereaved families.

Charles Moyer was disdainful.

"These are the same 'citizens' who said strikers were unfit to live in Calumet. You got along without them before; you can manage without them now. We can bury our dead. You must not accept a penny."

It wasn't easy for the traumatized families. Barely surviving . . . some had been forced to send women and children to live with working miners . . . now they faced burial expenses. They stood in doorways, nonetheless, courteous but firm when Citizens' Alliance volunteers knocked and offered envelopes of money. What Moyer did not know was that a handful of wives and mothers slipped to the back doors of the same volunteers late that night.

"We don't want Mr. Moyer or our men to know we're here. We had to refuse when you came to our houses," they said, weeping. "But there's no money to pay for the casket or the cemetery." Quick embraces; the envelopes changed hands.

With Annie, there was no halfway. She stood squarely behind her union leaders, and believed Mrs. Bert Czabo and Mrs. Anna Lustig who swore they had seen the white Citizens' Alliance button on the overcoat of the man who cried "Fire." Annie's sense of outrage burned with an intensity that surprised even her best friends. One of them said, after Annie brandished a broom at the backsides of two charity-bound volunteers, "Tall Annie's mad enough for ten people nowadays." In the heat of her anger, and probable self-blame, Annie truly believed unknown Alliance members had planned a cruel joke. She believed, too, later testimony of witnesses who said they heard gunshots outside and in the back regions of the hall before the panic.

Later, her passion cooled, Annie was to testify she saw neither the man who caused the tragedy nor the Alliance button he was allegedly wearing.

While Annie walked from one bereaved house to another that Christmas week, Moyer and his recently arrived colleague, Charles Tanner, hurried back to the Hancock Hotel where they had rooms. Tanner was the union's auditor and had been pressed into emergency service. The men were glum.

"We need to plan," Moyer said as they approached the hotel. Worry made him feel older than his 47 years. "With funerals scheduled for tomorrow, we need strategies to prevent any panic that might occur."

"Surely there will be guards," Tanner said.

Moyer nodded, "Of course." But Moyer knew there was always the gruesome possibility that shots would be fired. Certainly both the company men and strikers had been guilty of violence these past six months.

In the hotel lobby, Moyer and Tanner were surprised to see Sheriff Cruse approaching them. The sheriff introduced his partner as Mr. Peterman, the company attorney, and the four men ascended to Moyer's room. Once settled, Cruse came straight to the point.

"In a few minutes, a small delegation will be here to talk common sense," Cruse said, folding his arms across his huge midriff, on which hung a watch and chain. He was making an heroic effort to be amiable, but all present were uncomfortable. A knock sounded at the door and he stood up.

"That will be the committee," he said.

The group that filed in and introduced themselves were prominent members of the community. Some were also associated with the Citizens' Alliance.

Dr. Thometz, a dentist, spoke. "Mr. Moyer, the families of the victims have our every sympathy. We know they are facing bad times." He paused. "Why has financial relief been refused by these people? Won't you reconsider and accept this check?"

80

Moyer shook his head. "That's up to the strikers and the bereaved ones to decide, but if it were up to me, I'd not take a penny."

"You are responsible for their attitude," Peterman said. "You must rescind your orders."

"I've nothing to rescind," Moyer said, looking away from him.

Peterman went on. "You must also retract your statements that the Citizens' Alliance was responsible for the Italian Hall Tragedy."

Moyer held his ground, saying that a number of witnesses had told him that the man who cried "Fire" was an Alliance member.

"Well, sir, if you maintain this position," Peterman said with considerable heat, "You'll have to be responsible for your own safety."

Moyer stiffened. He felt the anger of the group and quickly agreed to go back to the Calumet union officers and try to downplay the rumor that the Citizens' Alliance was at fault.

The delegation left. Sheriff Cruse and Peterman took their coats and hats from the bed and departed.

Moyer had turned to pick up the tall, black telephone to ring union headquarters when the door swung open and several men, strangers, entered and moved toward Tanner. The union man stood rooted with surprise. The intruders called out for Moyer—Tanner saw they were unarmed, but his relief was brief. A quick look into the hall revealed that there were an unknown number out there, champing for action and wielding buggy whips and guns!

Moyer turned to confront the new "delegation," his heart thumping wildly. He saw fury fixed with determination on their faces and knew that he was about to be sacrificed.

•

CHAPTER 12

"A SHOT IN THE BACK
OF THE WORKING CLASS"

As a union organizer, Moyer had been in many hostile situations.

This, he decided was the ultimate. He tensed—at least he wouldn't go without a struggle. Moyer found himself being seized and dragged toward the open door. He fought back and a man rewarded his efforts with a blow from the handle of an automatic gun.

It discharged, sending a bullet into Moyer's back which lodged a few centimeters from his spine. He crumpled; his ears rang and pain enveloped his body. Someone shouted, "How bad is it?" Someone else laughed and said, "He ain't dead yet." At that same moment, someone struck Tanner; flesh on his face split open and blood welled.

Rough hands lifted Moyer and propelled both victims down the hall to the stairway and out to the street. The abductors now numbered more than a dozen and, as they dragged their prey out to the street, were met by a large, jeering crowd of Citizens' Alliance Members amassed in front of the hotel.

From there, the Tanner-Moyer situation went from bad to terrible. Individuals in the mob spelled each other with kicks, blows and threats of lynching and drowning in the Portage Canal.

It was more than a mile to the train depot in Houghton, but the crazed mob pushed and fought each other for the privilege of abuse. Bloody and dazed, Tanner was brought to full consciousness when a man hit him with brass knuckles, opening another wound, high on his cheek. Too hurt to care, neither Moyer or Tanner was aware that by now the station was crawling with men looking for action. Riot whistles began to blow.

Deputy Sheriff W.B. Hensley with a deputy then took charge. They had seen everything that had happened, starting at the hotel room and, according to Hensley's own reports, he had entreated the mob not to lynch the union men or throw them in the canal, but to allow him and his deputy to toss them on the night train to Chicago.

The officers wrested Moyer and Tanner from the mob's clutches, and pushed them onto the train. Hensley purchased tickets; saw that the victims washed themselves in the tiny compartment lavatory. Hensley could not help but feel like a hero. He summoned a porter.

"Give these men anything they request," he said. He indicated his deputy. "We're saving these men from a riot mob."

In the Pullman car, Moyer eased himself into a lower berth. Every inch of his body ached, and the wound pulsed miserably. He sensed he was in no danger of dying, but he wished heartily that sleep would come and blot out reality. Dimly, he was aware of Hensley's voice coming from the narrow doorway and with difficulty opened his swollen eyes. The deputy sheriff wore both a badge and a Citizens' Alliance button.

"Let's go over what happened from the beginning," Hensley said, impassively. He seated himself on the edge of the bunk with a paper and pencil.

Moyer sighed. Every movement hurt. Lying very still, he remembered Hensley. He was a hired strikebreaker, a gunman who had worked for western mine owners at a time when Moyer had been organizing a union there. He recalled, too, that Cruse had also used Hensley as a guard for union officials in Calumet.

In time, the two men agreed upon what the actual events of the past two hours had been, and Hensley rose and went to a sitting car. Tanner, his puffy face looking pale, held a telegraph form in front of his chief.

"Yes," Moyer sighed. "We need to let the boss in Denver know what happened."

Tanner took the dictation which ended with *"have the press say for me that the cause I represent is well worth the suffering I have undergone. The cause of the striking miners is just and they will win."*

Exhausted, he lay back in the berth. Hensley gave orders to the conductor that the message not be sent until he and his deputy had left the train.

Moyer, reasonably enough, had requested medical treatment. When the train stopped at a small town a few miles south of Houghton, Hensley asked a station agent to fetch help. Moments later, a doctor climbed aboard.

"Send the bill to Sheriff Cruse," Hensley said.

The train jerked to a start.

* * *

Annie heard the tale of abduction the next day. As heavy as her heart lay with grief for the families of the dead, she was incensed.

"Joe," she stormed, "they can't get away with this! Where is the freedom of speech we're guaranteed in this country?" She slammed the sad iron on the kitchen stove to re-heat. I should be doing something—anything, she thought, not here ironing shirts.

"Well," Joe said, "what can anybody do? Looks like our law enforcement men encouraged it. Hensley stayed on the train for awhile, then he got off and left two guards. And he arranged for a doctor, didn't he?"

"Very thoughtful." Annie's sarcasm showed a little of the intensity of her feeling. "Charles could have been killed. We still don't know how serious his injuries are."

"Maybe some of the fellas at the bar will have news," Joseph said.

He pulled a stocking cap over his ears and buttoned his old wool mackinaw.

* * *

To Moyer, the rail trip south was interminable. His wound had stopped bleeding, but pain pushed away all attempts to sleep. He was immensely grateful for Tanner's presence; the man had kept a cool head. Tanner stood, now, uncertainly, at the door of the sleeping compartment.

"Can I do anything to make you more comfortable?"

Moyer cautiously shifted his weight. "No. There's not much anyone can do until the bullet is cut out. How are you feeling, Charlie?"

"Passable. I ache all over."

Moyer managed a small grin. "I know the feeling. Where's Hensley now?"

"The two of them got off at the border. Now you try to rest."

*   *   *

A bevy of reporters interviewed Moyer at Milwaukee station. He recounted his abuse in Hancock, adding that at the last minute a tall, well-dressed, smooth-shaven man had alighted from an automobile, leaped on Moyer and tried to strangle him, threatening at the same time, to kill him. Suddenly releasing him, the attacker reached into Moyers overcoat, pulled out a wallet which he kept, along with thirty dollars it contained. At the onset, Moyer said he heard a voice shout,"There goes MacNaughton." In spite of the time Moyer had been in the Keewanaw Peninsula, he had never seen the man and was willing to believe that he was, indeed, being throttled by the manager himself.

Later, when it was proven MacNaughton could not possibly have been in the mob, Moyer had to admit his accusation was false. The choking incident, then, still remains a mystery. There was, too, the statement of Moyer's lawyer (not Darrow) from the hospital room, that Moyer and MacNaughton had been in close negotiation during the strike. This was a lie and Moyer lost credibility.

*   *   *

The next day, when the train steamed into the Chicago station, another crowd of reporters surrounded him. They saw the W.F.M. president trundled off the train in a wheelchair, his head and upper body bound with bandages, but well able to deliver a spirited criticism of his attack. He dwelt at length on the terrible plight of the mine workers.

"I'm going back in two days," he declared. "I will obtain a decent wage and working conditions for these men."

Federation friends then took the rumpled zealot to a hotel. Charles Tanner stayed close by his boss and in a few days Moyer entered St. Luke's Hospital for surgical removal of the bullet. He recovered rapidly.

*　　*　　*

The snows finally came. Lightly at first, it was a skittering that blew about the towering mine buildings, coated a thin frosting on the tailing piles and settled on the rooftops.

"Mama! It's me, Ana." She stamped slush from her shoes on the rag rug inside the front door. Untying her gray babushka, she walked to the kitchen of her parent's rented house. It was not where she had lived as a child. They had moved many times during those years.

"I came to see how Frankie's ankle is." Annie kissed her mother's cheek and sat down at the large wooden table spread with a red checked cloth.

"I wrapped it tight, and the swelling has gone down," Mary Klobuchar said. "He's getting around all right. You should see him playing with his Christmas toys."

Annie laughed. "Good." Her face turned serious. "Mama, I've posted a letter to Papa and Mary, telling them we're fine. I expect she'll write back quick."

Mary's mouth twisted in a wry smile.

"Well, since you know Pa's not much of a talker, I don't much expect him to write."

"You've got us kids, Ma. We're safe and well. A lot better off than some of our friends."

"I know, Annie. There's going to be plenty of helping out to do." She lay down a paring knife and potato and stared out the window. Snow fell fast and thick.

86

"It'll be a long, cold walk to the cemetery," she said.

*   *   *

Pewter gray and snowy, it was the last Sunday in December. Mourners left their church services, met others at street intersections, and with horse-drawn hearses moving one behind the other, the procession gathered to itself the small, white caskets and the larger, dark wood coffins. Perhaps fifty thousand pairs of feet trudged the mile-and-a-half down Pine Street to the Lakeview Cemetery.

Annie Clemenc, wearing her special, white dress, grasped the handmade flag, which she had draped with black crepe, and led the slow, burial march. How different from the many smiling, spirited parades she had headed.

A friend had once asked, "Isn't that flag and staff too heavy for you, Annie?"

She had answered, *"I get used to it. I carried it ten miles one morning. The men wouldn't let me carry it any further. I love to carry it."*

Now, the flagstaff wrenched at her shoulders and the once bright flame of determination only flickered in her heart. Tears stung her cheeks in the icy air. She walked through the cemetery gate, over a rise to where strikers themselves had dug into the rocky, frozen soil to open five trenches . . . three for Catholics and two for Protestants. A half-dozen single graves also awaited.

Annie lowered the flagstaff into a holder and stood rigidly, looking away from the gaping holes which were lined at the perimeter with wide boards. She counted fourteen hearses, three flatbed death wagons and an auto-truck as they drew up to the burial site. She heard drums beating slowly and music—a dirge floating eerily on the wind. Five hundred iron workers from Negaunee appeared in a long, wide column and, in time, the strikers came. Many carried flags and union banners and all wore fronds of the tamarack tree pinned to their lapels. Last came the two columns of coffin-bearers;the shoulders of four men supporting each burden. The procession encircled the graveside, lowered the caskets, and the eulogy began. Annie was never again to weep so bitterly.

The outpouring of grief from the mourners could be heard all the way back to Redjacket.

By nightfall, the trenches were closed, the mourners had retraced their steps and the cemetery lay guarded by scattered, leafless hardwoods and clusters of spire-like spruce trees.

*   *   *

An inquest had to be held. There were too many mysteries connected with the circumstances of the Italian Hall Tragedy, one of the worst in Michigan's history.

Village officials scheduled the inquest to take place in the Italian Hall, believing that the chosen sixty-nine witnesses could better describe and dramatize their testimony from there. Anthony Lucas, an attorney, and William Fisher, the coroner, were picked as examiners. Six jurors who were fairly representative of the population also were allowed questions.

Annie Clemenc testified that she had not seen a Citizens' Alliance button on the man who cried "Fire.".. *"I just heard a loud voice and all the kids started to holler 'Fire'. I just heard a man's voice holler 'Fire!'"*

"Was he apart from the rest or with a bunch of them?" Lucas asked.

"Could not tell you that. I was talking to the children, there . . . just heard a cry of 'fire' from the door. It was packed with men, women and children."

The Italian Hall inquest ended on December 31. During this time, the court recorder had set down hundreds of thousands of words from witnesses. The jury, with exceptional attentiveness, listened to the awful question, "Why was the fire alarm raised inside the hall?"—a question that was not answered in the verdict.

But there was evidence that a fire had been seen in one of the chimneys of the building. A man who lived close by had noticed sparks and smoke coming from the large hall and had phoned the Redjacket Fire Department. Several persons testified that they had seen a young boy descending a ladder at the rear of the hall before the fire whistle sounded. They said that the child's cap was glowing with sparks; others said there were embers on his shoulders. A man, they said, had lifted the lad down from a low shed and quickly smothered the fire on his clothing. This evidence, it would seem, points to a probable chimney fire. It became more credible when witnesses inside said that there was a flurry of activity before the cry of "Fire," by party goers looking out a second floor window after which a small group of men hurriedly left the hall. None of the latter citizens came forward to corroborate this testimony.

Veteran firemen later were quoted as saying a chimney fire would not have been serious.

In the end, the coroner's jury was unable to unravel enough corresponding facts to lay the blame on anyone. The man who cried "Fire!" has gone, probably, to his own grave, suffering the torments of the damned. If others knew his identity, they also hid the ghastly truth. The seventy-three death certificates recorded the cause as . . . "a jam through a false alarm call of fire by someone at present unknown."

*   *   *

The old year slipped toward 1914 but there was little merriment or ringing in the new as Calumet lay buried in its grief and heavy snows. Annie went about her household chores, cooking and cleaning for the renters in her large frame house on Pine Street, but she felt as depressed as those who had lost a loved one. Too, there was little to be optimistic about in regard to union-management relations. The companies had imported more miners and production was normal. Strikers left by the dozens every day, boarding the train for southern Michigan, hoping to hire into automobile factories.

Joseph was rarely home except to sit down for meals and roll into bed late from the tavern. They hardly spoke anymore.

"All this writing in the newspapers about you, Annie," he said. "You'll be getting a big head. And why do you bother to go to union meetings? There's not enough strikers left to make much difference. We've lost this battle."

"Don't say that!" Annie screamed. She threw a dish-towel at a hook on the wall and missed. She picked it up; stuffed it in a laundry basket.

"We must plan better strategies—stay closer in contact with the federation out west. We must hang on for the generations of miners coming after us." She warmed to the subject. "Look at Mother Jones—see what she has done by organizing and speaking up. She had tragedy in her life too. She never gives in to management."

"Sure. What does it get her? Locked up, that's what. I heard she just got out of jail in Colorado."

"Just what I've got to look forward to in a couple of weeks," Annie said, her shoulders slumping. "I wonder if I will get a choice of a fine or a jail sentence."

"Doesn't much matter," Joseph said. "I haven't got money to pay it. Maybe your boyfriend, Shavs, can bail you out."

"Leave him out of it!"

"Fine! I have to put up with everyone in town laughing behind their hand at me because of you carrying on with that little shrimp! I lose my wife and home, and you want me to leave out the cause of it!" Joseph stormed from the room. Annie stared after him a moment, cradled her head in her arms on the table, and cried.

Two weeks later, as scheduled, Annie traveled to Houghton where Judge O'Brian sentenced her to ten days in the county jail. Frank Shavs visited; her husband did not.

*   *   *

Once more in her own home, those deep, bitter January days, Annie fell ill. She had returned to an empty house; Joseph had left—the boarders found other accommodations in her absence. Annie lay feverish and dazed for several days until it because apparent that she needed more constant

90

attention. Mary Klobuchar and friends wrapped Annie in quilts and moved her by carriage to her parents' house. There, Mary, ever the nurse, tenderly cared for her eldest. Sometimes she had to call for help when Annie convulsed; during one attack it took five people to restrain her.

* * *

A grand jury, investigating Moyer's complaint of abduction, assault and battery, found no one guilty of illegal behavior toward him. The jury did find Moyer and more than three dozen W.F.M. officers and members guilty of conspiracy to prevent workers from entering Keeweenaw to look for employment.

Testifying in his own behalf, Moyer said *"When the shot was fired into my back, it was not into Moyer alone, but that shot reached every miner in the continent. It was a shot in the back of the working class and especially into the backs of organized labor."*

* * *

Upton Sinclair was pleased with the quick legislation reforms his exposé of the meat packing industry had triggered. He continued to be a lively socialist.

Now the morning newspapers fed him daily information about the copper miners' strike. The day Moyer's abduction was reported, Sinclair could contain himself no longer. He wrote President Wilson that the attack on Moyer was an outrage; he asked that the President order federal troops to guard Moyer's journey back to Calumet.

* * *

As she grew stronger, Annie's friends kept her abreast of union happenings. In early February, she walked home to a cold, empty house.

Annie knew in her heart, that neither she nor Joseph had tried hard enough to understand each other during the past stressful year. The marriage had failed. She would have to look ahead.

Frank Shavs' attentions and the plans Ella Bloor was formulating made the transition easier.

"My reporting here is finished," Frank said, "I can't foresee anything newsworthy breaking."

"I know," Annie said, "The company's operating with a full working force. But that doesn't mean we're giving up. We need money to hang on, that's true, but Ella and I are scheduling a speaking tour to raise funds. We'll bring in enough so the union miners who are left won't have to leave town."

"Good luck, then, Annie," Frank said, "I'll be on the morning train to Chicago. Promise you won't forget me!"

"How could I forget, you, Frank?" Annie laughed. "Every old wife in Redjacket has let me know their disapproval of the way I behave."

"Come to Chicago with me, Annie," he said, grasping her hand. "We can be married there." His dark eyes glowed.

"I want to be with you, too, Frank," she said, withdrawing her hand, "But I have a lot to do, personally and for the union. I'll write you often."

*   *   *

Ella Bloor's experience as a journalist and reporter for the Nationalist Socialist Press Association made it easy for her to map out a speaking tour. Their goal was to interest other miners in the strikers' fight by luring them to union halls with the prospect of meeting two women prominent in the labor struggle. Naturally, someone would pass a hat.

Ella designed a tour that would take them to Wisconsin, Illinois, Indiana, Ohio, Pennsylvania, West Virginia and Washington, D.C.

While Annie and Ella Bloor worked on schedules and train times, a Congressional investigation was being whipped together in Washington, D.C. at the request of John Hilton, lawyer for the W.F.M. and Congressman John McDonald. The two men had gone personally to plead.

President Wilson listened, wishing heartily that both labor and management would give a little. When they left, he turned to the European news releases on his desk. Everything pointed to imminent war, with the United States a possible entrant. It was even more depressing than the labor struggle.

Hilton and McDonald had been persuasive; the President agreed that an outside committee might better originate possible negotiable ideas.

Within days, President Wilson had appointed five federal legislators who titled themselves the "Court of Investigation" and left immediately for Calumet. (The President did not send an escort for Moyer.)

CHAPTER 13

A DREAM REALIZED

At the hearings in the copper country, Annie came forward to testify her part in the early clash of militias, deputies and parading miners.

Ella Bloor was assigned a table in the courtroom during the investigation. She wrote articles describing the daily testimonies and wired them to Socialist Headquarters.

One morning during the hearings, Congressman Taylor of Arkansas leaned across his table and spoke to Ella.

"Did you know that there are investigators here talking about going down into the mines to look things over?" he asked incredulously. "I wouldn't risk my life that way. I've got a family."

Ella did not try to hide her disdain. "Miners go down every day," she said. "They have families." She later wrote into her dispatches that some of the committee members did indeed descend to the pits. Taylor stayed behind.

One of the most interesting reports Ella turned out was about a Turkish-American who travelled from Minneapolis to testify.

"Why have you come here?" the mine owner's counsel asked him after he was sworn in.

"I heard there was a federal government investigation here and I made up my mind to come and tell my story and see what your government would do about it," he said, in heavily accented English. *"If you do not pay attention to my story, then I'll know what freedom means in this country."*

"I am a skilled machinist," he continued, *"but out of work. Copper Country Mine Agents were recruiting workers in Wisconsin and told me I was needed in Michigan as a machinist. When I asked about the strike, the agents said it was over, and then they told me to meet them at the train station in Minneapolis."* Ella glanced about the courtroom, satisfied that all eyes were riveted on the Turk. Her reporter's instincts told her that this immigrant, with his sense of independence and dignity, could blow the whistle on the shady practices of mine management. But her experience answered, saying that probably nothing would change.

"When I walked up to the train that was waiting at the depot, I saw that it was filled with workingmen, like me. I got into the coach, and found a seat. The men were grumbling and their faces were dark. I didn't feel so good myself, because I hear them saying that the strike is still on. Then the fella sitting beside me said, 'I think we're going to be used as scabs,' and before I can do anything the train started to move. We all waited until the train made a stop, and some of us got up and went to the door. But there was a man standing there, with a gun pointed at us."

The Turk shifted uneasily in his seat at the memory. "The fella said, 'We are just going for sandwiches,' and the guard said, 'No, you ain't' and he waved the gun at us. All the way to Calumet, we were prisoners.

When we got off the train, there were more men with bigger guns. They forced us to walk to the Ahmeek Mine, and the next thing I know, all of us got tools and start work as ordinary miners. Me—a good machinist!

When I come up after my first shift, I have to find a place to live. That was easy—lots of boarding houses. So I decided to stay until payday, after all, it was better than no work."

The questioning counsel rose. "Sir, in spite of the way you allege to have been brought here, you were treated decently, weren't you?"

"Decently!" exploded the Turk. "It was terrible down there in the mine. We were treated like criminals! And when payday came, I go up to the table for my check, the man told me I owe it all for train fare and job tools."

"What did you do then?" one of the investigators said.

"I told my landlady and she said that's nothing new. I can pay her later. Then I have to work and save for two more months before I have enough money to go home." He shrugged and spread his hands. "By that time the mines don't need us so much anymore, I suppose."

Records do not show that mine owners were ever made to answer these charges. Ella, as she went to the telegraph office with the story, somehow knew that would come to pass.

Other miners told of conditions underground that had changed very little in forty years. The temperature in Tamarack mine was a steady 100 degrees and the air was thick with gas and smoke. The men could bear to wear no more than heavy boots, pants, and perhaps a cotton undershirt to absorb their sweat. This particular mine management was the only one to send drinking water down to the men. There were but the crudest of toilets, poor ventilation and always, the miners said, the fear of sand chunks falling off and suffocating them. It was well known that trammers, men who filled and pushed the railway cars by hand, buckled under the inhuman treatment within seven years and had to quit, their health broken. When they complained on the job, the captain fired them.

"Mules are treated better than trammers," one miner said. "Trammers get the lowest pay and are worked the hardest," he went on. "We can't even afford to buy gloves— they cost two week's pay! And most of the rocks have to be lifted because they are too big for our shovels."

Miners generally agreed that their captains cheated them, especially those workers who did not speak English. All miners had problems, though, with broken promises, lies, extortion. The testimonies became more and more disheartening, with the mine management counsel, refuting the charges as exaggerations. "If these things were once true," they said, "they are no longer."

Yet when Finnish trammers, that spring of 1913, refused to fill and push twenty cars of ore each shift (knowing that other trammers' quota was only sixteen), they lost their leader. He had gone to the captain to complain and was fired. His co-workers quit in protest, but had to ask for reinstatement when their children cried for food. Old conditions prevailed. Complaints were met the same way . . . the spokesman was fired and again his loyal colleagues walked off the job. This time the company relented. Sixteen loads a day would be the new quota.

Trammers were exuberant. "It shows what we can do if we stick together," they said, slapping each other's bare backs. "Maybe the W.F.M. can help us. Maybe we should join and see what happens."

Much testimony in the investigation centered around the one-man drill which the company was introducing to cut cost and increase output. From the standpoint of the men using the new tool, it was a "widowmaker." "Usually sooner than later," one miner quipped.

The two hundred eighty pound drill and accessories were supported by a tall, imbedded iron post. Should a post be loosened or struck by debris or collapse from any cause, the driller was doomed. The nearest man would be five hundred feet away, and by the time the crushed miner was brought to the surface, from perhaps six thousand feet below, he was either dead or maimed for life.

There existed an aid fund for workers to which both company and miners contributed. The fee was deducted from their wage. It paid a small benefit, perhaps fifteen dollars a month to survivors and a token amount for disabled miners. There was a manufacturing company in town which hired crippled miners.

"Be careful, Einar, (or Giovanni, or Angus, or Peter, or Bloise, or Anton, as their origin might be,) or you'll end up making brooms!" the miners would warn new employees.

The company also had a pension program but most miners chose to join ethnic societies. Mine management could and did evade the issue of death and disability by claiming the present-day common law. This law specified that an employer could say the worker had been negligent or had willingly assumed the risks, or that a fellow worker was at fault.

Another argument to retain the two-man drill was that men could take turns walking back to the mine shaft to breathe fresh air. Ventilation was worse at the ends of the drifts and, if a worker did not remember to leave his drill periodically and take in fresh air, he might fall unconscious. On a one-man drill, a worker in trouble had no partner to help. Some mines did not provide first aid underground.

*   *   *

"Of course, we would rather die than give up," Annie said.

She stood with Ella, towering over all gathered at the depot. Friends and many of the parents of children killed in the Christmas Tragedy were there, waving and smiling through their pain. The South Slav Committee, with pride, presented Annie with a beautifully tailored black suit and a large velvet hat, also black, which had tiny pink feathers gracing the crown. Annie was greatly touched, and stroked the fine wool of the jacket as she thanked them. The first grand clothing of her life!

Annie and Ella settled themselves on the train; travelled west, then south and east speaking as scheduled and wherever they could fit in time at a free hall. It was a dream come true. She—a miner's daughter, travelling with the famous Mother Bloor, speaking for a cause that would be bound to better the quality of life for thousands.

In Chicago, their first night there, Ella warned her friend beforehand.

"Now, Annie, these teamsters are active—we're expecting about twenty-five hundred. So talk to them just like you talk to your fellow union members back in Calumet. But whatever you do, don't say anything about 'scab priests.' These men are mostly Catholic."

"I'll try, Mother," (this name was popular for women taking leadership roles in reform) *"But I can't help it. It makes me so mad when I think of priests trying to make scabs out of our men."*

Annie had no more than finished her opening remarks when she attacked the "scab priests." Ella flushed and clenched her hands. But the men only laughed.

In Chicago, Ella and Annie were entertained by wealthy William Bross Lloyd. He greeted them at the door of his elegant mansion near Lake Michigan with a hearty handshake. *"Ella!"* he boomed. *"Are you still alive! I thought you would be teaching the devil how to manage hell by this time!"*

Annie gazed about the palace-like home with astonishment. To think these rich people with servants and rooms full of books should be concerned about her and labor problems.

"And this is the famous 'Tall Annie!' Well, Annie, welcome to Chicago. You'll find you have stout friends here!"

A short time later, Ella explained.

"Williams' home is always open to me and anyone I bring along. He knows they are strikers pleading a cause. We may stay as long as we wish."

They finished unpacking and Annie sat on the silk bedspread. She drank in the beauty of the furnishings and was glad she had worn her new outfit.

At lunch time, Mr. Lloyd took the women out to his favorite restaurant.

As they left their parked auto, they discovered that the workers were striking. The pickets, all women, did not act energetic enough to suit Annie. She excused herself and strode to face the strikers.

"Girls!" she admonished. *"That's not the way to picket! Make a noise. Call out 'Strike on' or 'Strike' or ask the people not to go in."*

"I can see you are a zealot!" Mr. Lloyd chuckled. "Come on, ladies, let's go to another restaurant and talk this over."

At Lloyd's mansion, sometime in late afternoon, Ella noticed that Annie was nowhere around.

"I didn't see her go out, either," Lloyd said, and as the day faded to twilight, they fretted.

"I don't like this," Lloyd said. "Those everlasting mine company gents might have done her harm." He paced about the huge living room, with its gleaming dark furniture. Suddenly, he left. Ella waited, apprehensive, until he hurried in a few minutes later.

"I've sent scouts downtown—so let's not worry. They'll find her."

Ella looked pensive. "William," she said clapping her hands, "I'll bet she's with the striking restaurant girls."

Lloyd stared at her a moment, then as one they reached for their hats.

*　　*　　*

There was Annie, carrying a protest sign and almost daring citizens to cross the restaurant picket line.

"Oh, Annie, you had us so worried!" Ella called as Annie swept past her on the sidewalk.

"Don't concern yourself," Annie smiled over her shoulder. "Go have a cup of tea and I'll join you pretty soon."

CHAPTER 14

A NEW LIFE

Train-weary, Annie and Ella said goodbye and tearfully promised each other that they would keep in touch. Annie returned to her hollow boarding house on Pine Street. The speaking tour had been a new and exciting experience. She had traveled to places she never would have seen otherwise. Especially, Annie valued her friendship with Mother Bloor that deepened during the weeks on the road. But Annie was exhausted and not a little disappointed that their efforts had brought in no funds above their expenses.

During the long hours between cities, she had time to think of a future without Joe Clemenc. Separation and divorce were greatly against Annie's church beliefs; that her marriage had failed wrenched at her heart. Guilt spread through her body at times until she felt she might fall ill again with whatever fever she had in January. Other times, when she thought about fiery Frank Shavs and his passion for her and her causes, Annie's sense of wrongdoing faded. There was much to do and, as emotionally twisted as she was, Annie knew she had to get on with life.

As the week passed, Annie realized there was little she could do to further the efforts of strikers. Survival funds were running low. The man in charge of relief payments cut off money going to farmer-miners. Strikers who were managing because they grew their own food quickly asked for their jobs back. Social workers were alarmed when they discovered the few hundred remaining strikers' families were in terrible need. Annie winced as she recalled reports of their having no clothing—not even shoes or underwear and too little food to survive. And the winter temperature often twenty degrees below zero for days at a time.

* * *

Moyer was having trouble convincing the local strikers that the bills piling up would have to be paid. The union in the Keewenaw Peninsula was thousand of dollars in debt from loans to the Federation and other unions.

Mine managers had hired others until all but one mine were working at full capacity. Moyer could see the end . . . strikers would have to give in on some of their demands. He sent a delegation of six spokesmen to James Mac-Naughton's office on March 27, 1914. Surprisingly, the manager agreed to see them. The six representatives arranged themselves before the powerful man and Jim Elliott spoke.

"We have been sent to see whether the men formerly employed . . . can come back as they went out, without bad feelings toward any man, and according to your statement that you would shorten the shift and grant a minimum wage."

MacNaughton listened impassively, looked from one delegate to another as he spoke.

"At the present time practically every mine in our management is full handed," he said. *"We will not discharge one man from our employ from any of our mines for the sake of making an opening for a returning striker, nor will we re-employ any striker who does not first surrender his Western Federation of Miners' card."*

McNaughton saw the men's faces harden. He looked out the window of his office and went on.

"We will give former employees preference when hiring is necessary, but the company reserves the right to refuse to hire such of our former employees who have been arrested or convicted of violence or men who have incited to violence."

"What about the old company policy of not hiring men over forty?" Elliott asked.

"That alone would not bar employment," MacNaughton answered.

The men left his office feeling as powerless as when they entered.

On April 4, Moyer told his union members that the United Mine Workers of America had ended their assistance. W.F.M. locals in other states were complaining about the taxes they were paying to support the Michigan strikers. There had to be a cut in relief, he said. Reduced benefits began within days . . . two dollars and twenty five cents a week total for a single man . . . and up to only six dollars a week for a man with six or more children. It was an unlivable sum. Those residing in company houses would be evicted and places must be found for them.

Many meetings and speeches later, on Easter Sunday, the issue came to a sad conclusion. This April 12, strikers voted to end the struggle. It was not a majority vote in the Calumet mines area, however, but other mines—the Ahmeek, the Hancock and the South Range swung the vote. There was not enough faith that W.F.M. was the proper organization to win reforms.

Five days before the crucial decision Annie stood before Judge O'Brien at the Houghton County Courthouse and received a sentence . . . seventy-five dollars costs and fines for two convictions of assault and battery.

Such a lot of money, Annie thought as she boarded the electric train for Calumet. Well, it wasn't all in vain. She nodded to herself as she sat down at a window seat in the tram. We already got an eight hour day and better procedures for taking grievances to the bosses.

Her mind turned to the cleaning and packing chores awaiting her at the big frame house. It was time she left the north country.

*   *   *

Within days, Annie had embraced the last friend and

relative. She visited George's grave and carefully packed his photograph in her steamer trunk.

"I'm spending a lot of time, lately, on trains, eh?" she smiled through the wetness of her lashes. Mary Klobuchar hovered close by Annie as they waited for the Chicago train. Although winter had left, spring hadn't arrived. The wind was biting and damp.

"I'll write, Mama. I'll be just fine, you'll see. When Joe and Frankie get home from school, tell them I love them. And write me when Papa says he's coming back."

"Sure, Ana, sure. God go with you."

After numerous legalities were taken care of, Annie Klobuchar Clemenc married Frank Shavs in Chicago. Her new husband's chief, writer-editor Ivan Molek, offered to rent the young couple part of his two-family house on 31st Street. There were four rooms each, upstairs and down, and Shavs settled in on the top floor.

Molek, ever tolerant, was aware of Frank Shavs' eccentricities. which sometimes took the form of going shoeless and wearing overalls at the office. Molek explained to surprised visitors that Shavs probably felt he could better identify with the laboring class in this attire.

"I humor him," Molek shrugged. *Shavs is one to attack his job with a certain pathological exaggeration.*

Annie slowly adjusted to big city life. There was hardly enough to keep her occupied, but it was a pleasant street and their landlord had planted trees in the front yard. They had a vegetable garden in the back; looking down on it that first summer made Annie a little less homesick for Michigan. She marketed, cleaned, did laundry and cleaned some more. Free time she spent preparing for the baby. Her new Chicago friends teased her when she told them.

"Twenty-six years old and having a first baby! Hoee! Better late than never!"

Baby Darwina was born the last day of October, a few weeks after World War Two had erupted in Europe. She was to be an only child, but before she started school, she had her uncle Joe for company. Ana's younger brother completed grammar school in Calumet, then came to Chicago for work. Ana took him in as a boarder. It was during these early years that little "Dottie," as Darwina was nicknamed, was injured so badly in an auto accident that her left arm was amputated. The Shavs family lived through worry-filled nights at the hospital and a long recovery for the dark-haired child. By the time Frankie Klobuchar had left school at 14 to follow Joe and live with his sister, also, Dottie was well and delighted to have two uncles to pay attention to her.

Annie and Frank Shavs lived in the upstairs apartment on 31st Street until 1923, when Frank decided that seventeen dollars a week was too much rent (Molek had raised the rent) and the Shavs moved to 5611 Campbell Street on the other side of Chicago. In the meantime, "Little" Frank worked at a machine company and in the Rand-McNally Building Printers before he ended up in Detroit.

Mary and George Klobuchar moved to Detroit in 1922.

Annie never quite recovered from her distress at Dottie's crippling injury. Even worse, Frank Shavs became an alcoholic and a mean drunk, besides. Annie fretted about Dottie's future and for many years worked at two jobs to save money for her daughter's security.

Annie sewed at the Paragon Hat Company on the day shift, then afterwards, for as much time as she could muster, trimmed hats for a man named Weber, who catered to the "Gold Coast" people in North Chicago.

Little Frank and Joe, when they lived with Annie, earned their keep by housecleaning each Saturday.

"I know now why she called the Calumet Jail 'dirty,'" Little Frank grumbled as his sister made him scrub the same floors twice in one day.

On week-ends, Annie loved to entertain neighbors and men and women she worked with. She took pains with her attire at these times . . . she knew what was fashionable and was able to dress well through the clever use of her sewing abilities.

Her husband took little notice of her and as time went on, alcohol became more important to him. Once, in a fit of disgust, Annie ran downstairs to the basement and drained his entire keg of wine down the sewer.

Annie coped as best she could through the years, and never gave up her political views. She remained an active Socialist. Once, reading that Ella Reeves Bloor was to speak in Chicago, Annie counted the minutes until the day came for her appearance. Annie went to the Socialist Party auditorium, heart pounding with anticipation at seeing her dear old friend, with whom she had lost contact since they parted in Calumet.

Ella began her speech; Annie sat enthralled.

Suddenly, overcome with emotion, Annie rose to her feet and ran down the aisle to the platform, crying out, "Oh, Mother, Mother, Mother!"

Mrs. Bloor turned in surprise and Annie threw her arms around her, weeping.

Afterwards, the two women talked for a long time. Ella Bloor, in her autobiography, wrote that Annie told her that Shavs had "not been very good to her," and that she, Annie, was supporting herself and her daughter. Annie returned home that evening, her heart singing. She could not know she was never to see Mother Bloor again.

In the 1920's, when Annie was raising her daughter, most of the Klobuchars left Calumet. Darwina grew to be a tall, beautiful woman—striking in her resemblance to her mother as a young woman. Parents George and Mary settled in Detroit and Mary lived only two years after the move. George died in 1931. Young Mary had a grown family in Aurora, Minnesota; by this time, Joe Klobuchar had gone to New Jersey and Little Frank had left Chicago and Annie to stay with his parents in Detroit.

Joe Clemenc had left the Upper Peninsula, too, for work downstate. He lived with a stepbrother in Detroit, then boarded with Mary and George Klobuchar when they took up residence in Detroit. They were openly fond of their former son-in-law. Little Frank did not see him again until an old friend, Dr. Kerr, phoned, saying that Joe Clemenc was ill and asking about Frankie Klobuchar. Frank immediately visited his former brother-in-law and was shocked at the state of his health. "He was just a shell of a man," Frank said. Joe Clemenc died within days of the visit.

Frank Shavs degenerated through the years, ending his career as a factory night watchman.

At this writing, 1987, "Little Frank Klobuchar" is alive, well and retired on a farm near Chelsea, Michigan. Darwina married and had two children. It was a stable marriage; Annie passed into old age grateful for that.

On a summer morning in 1956, in the same house she had moved into thirty-three years before, Ana Klobuchar Clemenc Shavs (or Shaws) died of cancer in her bed. She was sixty-eight years old. She was buried at Resurrection Cemetery in the north 1/2 section of lot 45. Her death certificate read: Occupation—Hatmaker.

Ana Clemenc Shavs was an instigator, parade flag bearer, and inspirational leader for the nine crucial months union copper miners struck.

She was a glorious display of fireworks in the dark skies of labor- management conflict. Her colors exploded briefly, yet remain brilliant in the history of Michigan's Upper Peninsula.

* * *

EPILOGUE

The Michigan Women's Studies Association in July, 1980, honored Annie's memory with the purchase of a strong, sensitive painting by the West Virginia artist, and member of the Miners' Art Group, Andy Willis. He portrayed Annie with her beloved flag, doing what she liked most—leading a march.

At the same time, Annie became the first person to be nominated for the Michigan Women's Hall of Fame and Historic Center.

Her youngest brother, Frank, attended the Lansing ceremony, appearing younger than his seventy-five years. He listened with pride and some sadness as an Executive Declaration from William Milliken, Governor of Michigan, was read proclaiming June 17, 1980, to be Annie Clemenc Day in Michigan. It urged all "citizens to familiarize themselves with the courage and resolve of this long-forgotten giant of Michigan labor history from the Upper Peninsula."

Michigan's Legislature also passed resolutions honoring Annie and commemorating her portrait.

Calumet citizens, in October of 1984, at length decided that the Italian Hall, long-abandoned, had become a possible safety hazard. Because of very high demolition costs, a volunteer force formed and the sturdy, memory-filled building fell to the wrecker's ball. Truckloads of the still usable brick were offered for twenty dollars a truckload.

Not long after, a Michigan playwright, John L. Beem, fascinated by the mystery of the Italian Hall Tragedy, wrote a play called "Mother Lode," which was produced in Ann Arbor.

References

American Great Crises in History. Vol IX and X. Americanization Dept. of Veterans of Foreign Wars of the U.S., Chicago, Illinois. 1925

Andrews, Clarence A. *'Big Annie' and the 1913 Michigan Strike.* Michigan History, 1957 (Spring 1973): 53–86

Bloor, Ella Reeve. *We Are Many.* New York, International Publishers. 1940

"The Annals of America." *Encyclopedia Britannica.* 1968. Vol. X, XI, XIII.

History of the Diocese of Sault Saint Marie and Marquette Village and Calumet, Michigan. Souvenir Centennial Book, 1875–1975

Klobuchar, Frank. Ana's only surviving close relative. Letters, photographs, family documents and oral recollections.

Miner's Bulletin, Calumet, Michigan. October 2, 1913, p.1.

Molek, Ivan (John). *Slovene Immigrant History, 1900–1950.* Translated by Mary Molek. Dover, Delaware. 1979.

Sullivan, William A. *The 1913 Revolt of the Michigan Copper Miners.* Michigan History 43 (Sept. 1959): 294–314

Thurner, Arthur W. *Calumet Copper and People: History of a Michigan Mining Community,* 1864–1970. Hancock, Michigan. 1974

Thurner, Arthur W. *Rebels on the Range: The Michigan Copper Miner's Strike of 1913–1914.* Lake Linden, Michigan. Forster Press. 1984

Fowke and Glazer, Eds. *Songs of Work and Freedom.* New York. 1960

Index